SpringerBriefs in Population Studies

Population Studies of Japan

W0235276

The world population is expected to expand by 39.4% to 9.6 billion in 2060 (UN World Population Prospects, revised 2010). Meanwhile, Japan is expected to see its population contract by nearly one third to 86.7 million, and its proportion of the elderly (65 years of age and over) will account for no less than 39.9% (National Institute of Population and Social Security Research in Japan, Population Projections for Japan 2012). Japan has entered the post-demographic transitional phase and will be the fastest-shrinking country in the world, followed by former Eastern bloc nations, leading other Asian countries that are experiencing drastic changes.

A declining population that is rapidly aging impacts a country's economic growth, labor market, pensions, taxation, health care, and housing. The social structure and geographical distribution in the country will drastically change, and short-term as well as long-term solutions for economic and social consequences of this trend will be required.

This series aims to draw attention to Japan's entering the post-demographic transition phase and to present cutting-edge research in Japanese population studies. It will include compact monographs under the editorial supervision of the Population Association of Japan (PAJ).

The PAJ was established in 1948 and organizes researchers with a wide range of interests in population studies of Japan. The major fields are (1) population structure and aging; (2) migration, urbanization, and distribution; (3) fertility; (4) mortality and morbidity; (5) nuptiality, family, and households; (6) labor force and unemployment; (7) population projection and population policy (including family planning); and (8) historical demography. Since 1978, the PAJ has been publishing the academic journal *Jinkogaku Kenkyu* (The Journal of Population Studies), in which most of the articles are written in Japanese.

Thus, the scope of this series spans the entire field of population issues in Japan, impacts on socioeconomic change, and implications for policy measures. It includes population aging, fertility and family formation, household structures, population health, mortality, human geography and regional population, and comparative studies with other countries.

This series will be of great interest to a wide range of researchers in other countries confronting a post-demographic transition stage, demographers, population geographers, sociologists, economists, political scientists, health researchers, and practitioners across a broad spectrum of social sciences.

Yoshitaka Ishikawa
Editor

Japanese Population Geographies II

Minority Populations and Future Prospects

 Springer

Editor
Yoshitaka Ishikawa
Professor Emeritus
Kyoto University
Kyoto, Japan

ISSN 2211-3215   ISSN 2211-3223 (electronic)
SpringerBriefs in Population Studies
ISSN 2198-2724   ISSN 2198-2732 (electronic)
Population Studies of Japan
ISBN 978-981-99-2075-4   ISBN 978-981-99-2076-1 (eBook)
https://doi.org/10.1007/978-981-99-2076-1

This Springer imprint is published by the registered company Springer Nature Singapore Pte Ltd.
The registered company address is: 152 Beach Road, #21-01/04 Gateway East, Singapore 189721,
Singapore

Preface

According to the World Population Prospects 2022 issued by the Population Division, United Nations, the population growth rate in Japan in 2021 was −0.54%. Among the 38 OECD countries, a negative rate has also been recorded for 11 countries (Czechia, Estonia, Greece, Hungary, Italy, Latvia, Lithuania, Poland, Portugal, Republic of Korea, and Slovakia); Japan's rate of decrease follows those of Latvia (−1.38%), Lithuania (−1.29%), and Greece (−0.69%). However, if only countries with a population of more than 50 million were counted, Japan would have the highest rate of decrease. Therefore, Japan is a representative population-decreasing country in the contemporary developed world. Furthermore, the COVID-19 pandemic from 2020 has caused a sudden decline in total fertility rates (TFR) throughout the world. If the pandemic further accelerates the decline of TFR, the era of depopulation will come sooner than expected. This means that a rapidly increasing number of countries will have to deal with the risk of population decline as an urgent issue.

Over the past decade, the term *jinko gensho* (population decline) has become a popular expression in Japan: The total population of Japan began to drop after reaching its peak (128.08 million) in 2008. Following the long-term population projections by Japan's National Institute of Population and Social Security Research (IPSS), showing that the country's annual decrease will gradually speed up, this issue's seriousness has become widely recognized, prompting myriad discussions on population decline.

It goes without saying that this demographic trend is a painful experience for Japan. Nevertheless, the recent population-related trends observed in Japan should provide many valuable lessons for countries around the world that are in the midst of population decline or facing its imminent emergence. Consequently, at this juncture we have an obligation to let other countries know about the achievements in population geographic studies in Japan. Existing research in Japan has received substantial stimuli from foreign research, especially from English-speaking countries. Unfortunately, however, Japanese population geographers are reluctant to publish their findings in English, often publishing their work only in Japanese journals. Thus, even papers with excellent findings and insights are not widely known around the world.

Therefore, the editor came up with the idea of compiling an anthology of English translations of recently published Japanese papers that are regarded as important achievements in Japanese population geography. Given the rapid progress of automatic translation in recent years, this may be the last chance to publish a fully translated anthology, and it's possible that similar projects will not be carried out in the future. It should be noted, however, that not all recent achievements in population geography in Japan deal with issues related to population decline, and interest in them varies. Accordingly, the relationship to population decline in each of the anthology's papers is explicitly described in its introduction below.

This work is published by the Population Association of Japan (PAJ) as a *Population Studies of Japan* Series through Springer. According to the agreement between PAJ and Springer, the length of a single book is limited to 125 pages, which is insufficient to also include the most recent five papers. For this reason, it was decided to publish a total of ten papers in two volumes, each containing five papers.

The two English volumes comprise the following original papers published in Japanese.

The first volume (Ishikawa Y (ed) (2023) *Japanese Population Geographies I: Migration, Urban Areas and a New Concept*) includes the following five chapters:

Chapter 1: Inoue T (2016) *Posuto jinko tenkan-ki no jinko ido* (Internal Migration in the Post-Demographic Transition Period). In: Sato R., Kaneko R. (eds) *Posuto jinko tenkan-ki no nihon* (Japan in the Period of Post-demographic Transition), Hara Shobo, Tokyo, 111–133.

Chapter 2: Ishikawa Y (2016) *Nihon no kokunai intai ido saiko* (Internal Retirement Migration in Japan Revisited). In Ishikawa Y. (ed) *Ryunyu gaikokujin to nihon: Jinko gensho no shohosen* (New Immigration and Japan: Solution to Population Decline), Kaisei-sha, Otsu, 119–146.

Chapter 3: Yamada H (2020) *Higashi-nihon daishinsai no hisaichi ni okeru kyojuchi ido to shigaichi saihen tono kankei: Tohoku chiho no hisai-ken ni chakumoku shite* (Characteristics of Residential Mobility After the Great East Japan Earthquake: Focusing on Affected Prefectures of the Tohoku Region, Japan). *Kikan Chirigaku* (Quarterly Journal of Geography) 72(2): 71–90. https://doi.org/10.5190/tga.72.2_71

Chapter 4: Kanda H, Isoda Y, Nakaya T (2020) *Jinko gensho kyokumen ni okeru nihon no toshi kozo no hensen* (Spatial-Cycle Model Phases and Differential Urbanization of Cities in the Era of National Population Decline: Japanese Cities 1980–2015). *Kikan Chirigaku* (Quarterly Journal of Geography) 72(2): 91–106. https://doi.org/10.5190/tga.72.2_91

Chapter 5: Sakuno H (2019) *Jinko gensho shakai ni okeru kankei jinko no igi to kanosei* (Significance and Possibilities of the New Concept of "Relationship Population" in Japan's Population Decline Society). *Keizai Chirigaku Nenpo* (Annals of the Association of Economic Geographers) 65(1): 10–28. https://doi.org/10.20592/jaeg.65.1_10

The second volume (Ishikawa Y (ed) (2023) *Japanese Population Geographies II: Minority Populations and Future Prospects*) includes the following five chapters:

Chapter 1: Takeshita S, Hanaoka K, Ishikawa Y (2020) *Hetero-rokarizumu-ron no kensho: Aichi ken no toruko-jin no kyoju patan ni shoten o atete* (Investigating

Empirical Validity of Heterolocalism: Focusing on Turkish Residential Patterns in Aichi Prefecture). *Aichi Gakuin Daigaku Bungakubu Kiyo* (Bulletin of the Faculty of Letters of Aichi Gakuin University) 50: 65–74

Chapter 2: Yamauchi M (2021) *Osaka-shi ni okeru seiteki mainoritei no kukan bunpu* (Examining Geographic Distribution of LGBTs in Osaka City, Japan). *Jinko Mondai Kenkyu* (Journal of Population Problems) 77(2): 185–205. https://www.ipss. go.jp/syoushika/bunken/data/pdf/21770207.pdf

Chapter 3: Koike S (2021) *Nihon no chiiki-betsu shorai jinko no mitooshi* (Future Prospects of Regional Population in Japan). *Jinko Mondai Kenkyu* (Journal of Population Problems) 77(2): 85–100. https://www.ipss.go.jp/syoushika/bunken/data/pdf/21770201.pdf

Chapter 4: Tanimoto R (2017) *Toshi kogai ni okeru byosho heno akuseshibiritei no shorai suikei: Osaka toshiken hokubu no jirei* (Future Projection of Accessibility to Hospital Beds in the Suburbs: The Case of the Northern Osaka Metropolitan Area). *Jimbun Chiri* (Japanese Journal of Human Geography) 69(4): 425–446. https://doi. org/10.4200/jjhg.69.04_425

Chapter 5: Nakazawa T (2018) *Seiji keizaiteki jinko chirigaku no kanosei: "Shukusho nihon no shogeki" o tegakari ni* (Toward a Politico-Economic Population Geography: A Critique of *The Shock of a Shrinking Japan*). *Keizai Chirigaku Nenpo* (Annals of the Association of Economic Geographers) 64(3):165–180. https:// doi.org/10.20592/jaeg.64.3_165

These 10 papers were originally published in Japanese, and for them to be included in this series, the editor had to ask the authors of each chapter to shorten them due to the length limitation mentioned above. Thus, several of the English translations are shorter than the original papers in Japanese, with some of them shorter by almost half. Consequently, in those substantially truncated English papers some of the excellent information from the original Japanese versions had to be omitted.

The following gives a brief overview of the significance of each paper in this series. The first volume focuses on internal migration, which has been the central theme of population geographic studies in Japan, as well as the structural changes in metropolitan areas, which are closely related to internal migration. It also includes a paper discussing a new population concept originating from rural areas in peripheral regions where population decline is most severe.

Inoue's paper in Chap. 1 analyzes the internal migration in post-war Japan and examines the influence of the demographic transition on so-called migration turnarounds in the post-demographic transition period. The author is one of Japan's leading population geographers and has many achievements in this field. Changes in migration between the three major metropolitan areas of Tokyo, Osaka, and Nagoya, which are the country's core regions, and their peripheral regions have been the focus of much attention in studies on population geography in Japan. The author discusses when and how the balance changed between the mainstream migration from the periphery to the core and the counterstream migration from the core to the periphery, as well the factors behind this change. Focusing on the migration turnarounds that have been observed twice in post-war Japan, the author carried out a wide-ranging study. The measure called "cohort cumulative social increase ratio," proposed by the author through his research, is particularly useful in studies that delve into the Japanese characteristics mentioned above.

Ishikawa's paper in Chap. 2 examines retirement migration, which has long attracted considerable attention in population geographic research in Japan. The predominant view is that retirement migration triggered by mandatory retirement has been widely observed in Western countries but not commonly in Japan. However, using the data from the 2010 Census, the author conducted a comprehensive analysis by creating a migration schedule for all prefectures and municipalities. Since retirement migration tends to be directed to rural areas where living costs are low, the influx of retired migrants is good news for municipalities in rural areas with marked population decline. From his analysis, the author confirms remarkable human flows to many prefectures/municipalities in peripheral Japan; this result indeed provides positive evidence of retirement migration. He also mentions the regional differences and assessments of retirement migration by officials of local governments in the major destinations of this migration. The paper mainly deals with the population group with a large cohort size, including the baby boomer generation born in the late 1940s, in which retirement migration has been particularly noticeable.

Yamada's paper in Chap. 3 is an important work that comprehensively clarifies the residential migration necessitated by the Great Eastern Japan Earthquake in 2011, based on the population migration data published in the 2015 Census. Although most internal migration in contemporary Japan has been voluntary, it is significant in that it is a typical form of displacement migration. The 2015 Census was originally planned to exclude the question on the respondent's address five years ago, which serves as the source of population migration data. However, in order to learn more about the damage caused by the earthquake, it was included among the survey items at the last minute. The author analyzed these data in detail and presented the results in several detailed maps, which is an important feature of the paper. The author argues that residential mobility in the affected areas can be viewed as a process carried out by migrants to recover their daily lives, and the different methods used to secure "safety" for each victim make this mobility complicated and diverse.

The paper of Kanda et al. in Chap. 4 reports on the transitions in the urban structures of 109 cities in Japan based on the constituent municipal population of their functional urban regions from 1980 to 2015. Since the 1980s, Japan's population has become more concentrated in the Tokyo metropolitan area, leading to a wider disparity with other regions. How to remedy this problem of mono-polar concentration in Tokyo has been an important concern in domestic regional policies over the last three decades. The authors examine how Japanese metropolitan areas have undergone structural changes in terms of whether the spatial-cycle model and the differential urbanization theory proposed based on Western experience apply to Japan. In conclusion, they claim that the disurbanization phase of the spatial-cycle model is not observed. Population decline occurred earlier in smaller cities than in the larger cities, reflecting the changes in urban structure, in contrast to the prediction of differential urbanization theory. This finding demonstrates the uniqueness of Japan's experience in relation to those of Europe and the USA.

Sakuno's paper in Chap. 5 discusses the significance and possibility of "relationship population," a new population concept developed in Japan. In contemporary Japan, the majority of municipalities outside of the Tokyo area are facing severe population decline, and there is a strong sense of exasperation in the regions. To overcome this situation and revitalize the regions, we need to view the population of a specific municipality not from the perspective of the resident population, as in the past, but from the relationship population. This approach in Japan of viewing a population in terms of the diverse relationships individuals have established with municipalities and regions other than their primary home has attracted wide attention. The author is a leading geographer conducting research on municipalities and communities in rural areas, where population decline and aging are progressing. On the basis of his previous works, the author examines in detail the various possible interpretations of the framework of this new population concept. It can be said that this paper was the result of efforts to comprehend the current situation and explore a future vision for local governments in rural areas where population decline is becoming a serious problem.

The second volume focuses on Japan's minority populations, including foreign residents and LGBT persons, as well as future visions for Japan and for specific regions and the important politico-economic perspective for studying the current/future Japanese population.

The paper of Takeshita et al. in Chap. 1 discusses the residential patterns of Turkish residents living in Aichi Prefecture. In Japan, interest in international migration and foreign residents has increased along with the progress of population decline. In particular, the formation of ethnic enclaves has attracted wide attention as a theme for geographic studies. The three authors have studied ethnic enclaves using a combination of population census microdata and participant observations. Enclaves of nationalities with a population of more than 100,000 are already well understood. Unfortunately, there has been little prior study on the formation of enclaves of nationalities with smaller populations. This chapter examines dispersed residential patterns from the standpoint of heterolocalism, using the case of Turkish residents. The authors believe that such a point of view may also be applicable to many other small nationality groups in Japan.

Yamauchi's paper in Chap. 2, in contrast to the previous chapter, is the result of studies on sexual minority populations, which have not been studied extensively by population geographers in Japan. The author, who has been studying the population geography of birth for many years, is the first to work on this minority population group. There is a certain volume of research accumulated in Europe and the USA in regard to LBGT persons, where they have been reported to have formed enclaves in large cities. This paper, which is a pioneering article on the population geography of LGBT persons in Japan, examines whether such enclaves can also be found in Japan, particularly in Osaka City. The author concludes that there is no significant association between the presence of LGBTs among respondents and distinct areas where LGBTs are found to be concentrated. Although it is not easy to clarify the geographic distribution of population groups with few members, it is noteworthy that this paper carefully elucidates their distribution by combining a few statistical methods. This is an analytical method that can be applied to studies of other minority groups in similar situations and provides many insights into them.

Koike's paper in Chap. 3 discusses the future of population distribution in Japan using data from regional population projections. In Japan, since the 1980s, the mono-polar concentration into Tokyo has continued, widening the disparity between the Tokyo area and the other regions. This has attracted much attention as one of the biggest problems in Japan, and considerable policy actions have been made to rectify it. It is important to note that the study is based on population projections of geographic units such as prefectures and municipalities conducted by IPSS, and data of projections up to 2045 have already been released. Furthermore, the author has played a central role in making the projections, and in this chapter, he uses these projection data and other sources to examine in detail the possible future population distribution. Interestingly, although he takes into account the increasing population trend of foreign residents in peripheral areas, the author concludes that the mono-polar concentration into Tokyo will continue in the future.

While Chap. 3 examines Japan's population distribution up to 2045 for the entire country, Tanimoto's paper in Chap. 4 projects accessibility to hospital beds up to 2025 for the northern part of the Osaka metropolitan area. IPSS has made projections for each municipality using only the single variable of population. Meanwhile, this paper is exemplary in that it has intentionally carried out regional projections in geographic units below the municipality level. It should be noted that such a detailed study was made possible by the work of Takashi Inoue, the author of Chap. 1 of Volume 1, on population projections at the small area (*chocho-aza*) level covering the entire country. In Japan, which is now in an era of depopulation, there is growing interest in proximity to various social services, including medical care at the community level, below the municipality level. This chapter is of great interest as a pioneering study on an increasingly important topic in Japanese population geography.

The next chapter, Chap. 5, draws attention to the importance of a politico-economic perspective in Japanese population geography. The author of this chapter has published many excellent papers on migration, focusing on life course at the individual level. The paper explores the possibility of establishing politico-economic population geographies by examining how actors in various regions present the problems associated with population decline, on the basis of "The Shock of a Shrinking Japan" published in 2017. In the past, Japanese geographers have been nearly oblivious to the political position of their research. In recent years, however, it has become necessary to discuss policy implications in the publication of results of population geographic studies. Nakazawa's paper is significant in that it compensates for the weaknesses of Japan's traditional population geographic studies and clarifies directions for the future by developing a more in-depth critical examination and discussing the importance of geopolitics in population research.

It is my sincere hope that this anthology of recent outstanding works in Japanese population geography will be read with interest by many people all over the world. Finally, we'd like to express our heartfelt thanks to the Japan Society for the Promotion of Science (JSPS), which made the research leading to the publication of this book possible (No. 21H00637). We are also indebted to Ron Read of the Osaka Branch of Human Global Communications Co., Ltd., for carefully and kindly editing the earlier English manuscripts.

Kyoto, Japan   Yoshitaka Ishikawa

Contents

Chapter 1
Investigating Empirical Validity of Heterolocalism: Focusing on Turkish Residential Patterns in Aichi Prefecture

Shuko Takeshita, Kazumasa Hanaoka, and Yoshitaka Ishikawa

Abstract The purpose of this study is to examine the residential patterns of ethnic minority Turks in Aichi Prefecture and to evaluate the applicability of heterolocalism as the framework for analyzing this issue. This work is conducted on the basis of insights gained from the microdata of the 2015 Population Census and participant observations. Residents from overseas in Japan represent 195 countries and regions, and of these points of origin, 171 account for populations of less than 10,000 within Japan. However, little is known about their residential patterns in Japan due to the very small number of existing studies on immigrant nationalities with small populations. Therefore, it is important to clarify their actual patterns of residence. The residential pattern of immigrant Turks in Aichi Prefecture are found to meet the five conditions of heterolocalism: (1) spatial dispersion of residence by immigrants of a particular nationality, (2) spatial disjuncture between home and work, (3) formation of an ethnic community without propinquity, (4) emergence of heterolocalism as a time-dependent phenomenon, and (5) formation of networks beyond the metropolis. Consequently, the heterolocal model of Zelinsky and Lee provides a good explanation of the residential patterns of Turks in Aichi Prefecture. In this view, networks of migrants who share religious and ethnic identities create heterolocal conditions on a variety of scales across regions and borders.

Keywords Heterolocalism · Residential pattern · Ethnic community · Ethnic minority · Turkish residents · Japan

S. Takeshita (✉)
Department of Japanese Cultural Studies, Aichi Gakuin University, Nissin, Aichi, Japan
e-mail: shtak@dpc.agu.ac.jp

K. Hanaoka
College of Letters, Ritsumeikan University, Kyoto, Japan

Y. Ishikawa
Kyoto University, Kyoto, Japan

1.1 Introduction

As of the end of 2019, there were 2,933,137 foreigners living in Japan representing 195 countries and regions. The number of foreign residents in Japan increased by 202,044 from the previous year to reach a record high (Immigration Service Agency of Japan 2020a). Of these, five nationalities have a resident population of over 100,000: China (813,675), South Korea (446,364), Vietnam (411,968), the Philippines (282,798), and Brazil (211,677). Much of the previous research on foreign residents in Japan have focused on these five major nationalities (e.g., Abe 2011; Yoshida 2011; Takahata 2012; Fukumoto 2013; Kataoka 2013; Yamashita 2019). An important achievement along this line of research was made by Ishikawa (2021), where the microdata of the 2015 Population Census were used to examine ethnic enclaves comprehensively and in detail, significantly advancing the study of ethnic enclaves in Japan. However, Ishikawa (2021: 187) pointed out the lack of studies on the living situations of foreign nationalities representing populations of less than 100,000.

There are, in fact, 171 countries and regions that represent resident populations in Japan of less than 10,000, and these account for 88% of the total number of countries/regions representing foreign residents in Japan (Immigration Service Agency of Japan 2020b). However, due to the very small number of previous studies on nationalities with small populations (Ishikawa and Hanaoka 2021), little is known about their residential patterns in Japan. Therefore, it is important to look into their actual patterns.

Here, we focus on the Turkish people of Aichi Prefecture as an example. Chain migration by people from Ordu Province, Türkiye[1] began in Aichi Prefecture in the 1990s, leading to the formation of the Turkish community in the 2000s. Members of the community are held together by a combination of shared networks: They all emigrated from the same province in Türkiye, they all engage in the same type of occupations, and they share the same religion.

Another interest of this study regards which of the existing explanatory frameworks associated with enclave formation best explains the actual residential patterns of nationalities with small populations. In this context, Zelinsky and Lee (1998) not only criticized the spatial assimilation and pluralism frameworks that were widely accepted at the time but also proposed heterolocalism as an alternative model for the new situations faced by new immigrants.

According to this model, immigrants begin to rapidly disperse spatially after entering the host country but retain their ethnic identity through telecommunications and personal visits (Zelinsky and Lee 1998; Wright and Ellis 2000; Hardwick 2006). Heterolocalism is, therefore, considered a valid viewpoint for describing the social and spatial situations of new immigrant groups (Sugiura 2011: 44). Ishikawa (2021) has also inferred that the theory of heterolocalism would apply to many of the enclaves

---

[1] Turkey changed its country name in the United Nations to Türkiye on June 1, 2022.

of newcomer foreigners in Japan. However, there has been no detailed examination of this framework's validity in Japan.

Accordingly, this paper attempts to analyze the residential patterns of ethnic minority Turks in Aichi Prefecture and to examine the applicability of heterolocalism as the framework for this analysis. This study is conducted on the basis of insights gained from the microdata of the 2015 Population Census and participant observations of Turks living in Aichi Prefecture curried out by the authors since 1999.

1.2 Distribution of the Turkish Population

According to the 2015 Population Census, the Turkish population in Japan increased 12.8 times from 205 in 1990 to 2615 in 2015. By prefecture, the highest number of Japan's Turkish residents, or roughly a third, live in Aichi Prefecture (813), followed by Saitama Prefecture (526) and Tokyo Prefecture (452)[2] (Figs. 1.1 and 1.2). Furthermore, Aichi Prefecture is home to many immigrants from Ordu Province, while Saitama Prefecture hosts many Kurds of Turkish nationality.

The Turks in Aichi Prefecture totaled 813 (686 men and 127 women), pointing to an overwhelming majority of men at 84.4%. Throughout Japan, men made up 81.5% (1816) of the Turkish population (425 women). Their employment rates in Aichi Prefecture were 61.8% for men and 26.7% for women, while nationwide they were 59.8% for men and 21.9% for women, showing a markedly lower employment rate for women.

Comparing the Turkish residential patterns in Aichi Prefecture and in Saitama Prefecture, we see that while there is a large Turkish population in the northwest part of Aichi Prefecture (Fig. 1.3), there is no clear concentration of Turks such as that seen in Kawaguchi City in the southeastern part of Saitama Prefecture (Fig. 1.4).

---

[2] According to "Statistics on the Foreigners Registered in Japan," the Turkish population in Japan as of the end of 2019 was 5,419. By prefecture, Saitama Prefecture has the largest number at 1,627, followed by Aichi Prefecture (1,393) and Tokyo (890). The number of Turks in Saitama Prefecture has increased rapidly, surpassing Aichi Prefecture as the highest from 2018. There are differences in the number of foreigners in Japan between the two sources of the "Statistics on the Foreigners Registered in Japan" and the "Population Census." Possible reasons for this are discussed in some detail in Ishikawa (2005).

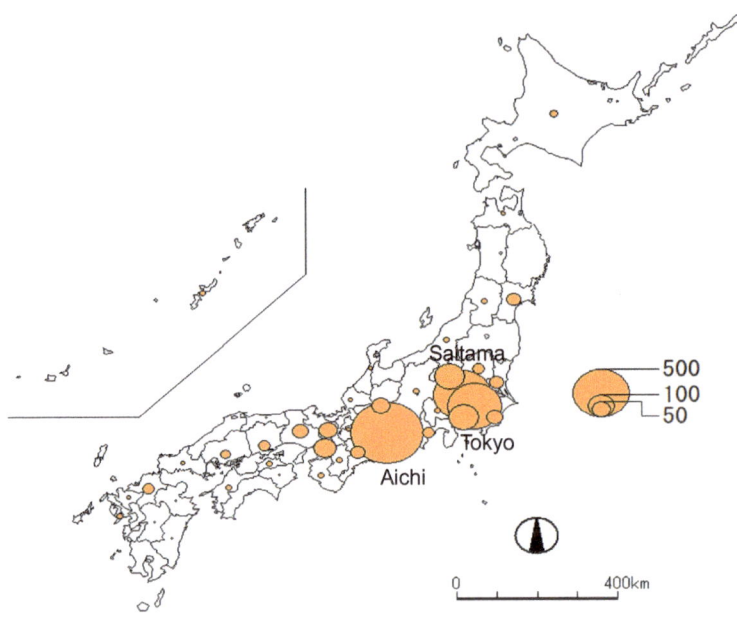

Fig. 1.1 Distribution of Turkish residents in Japan. *Source* Microdata of the 2015 Population Census

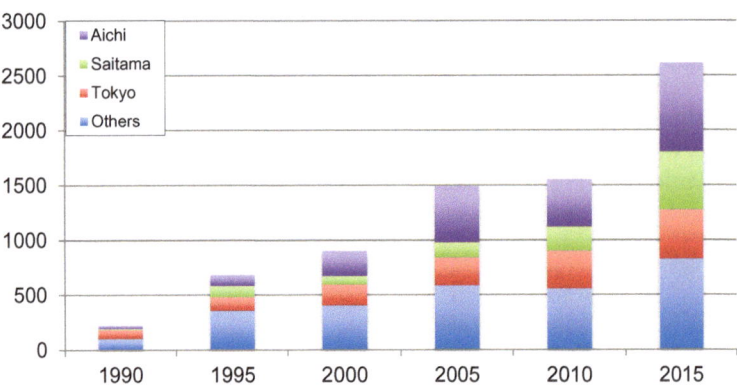

Fig. 1.2 Turkish population in Japan. *Source* 1990–2000 Population Census and 2005–2015 Microdata of Population Census

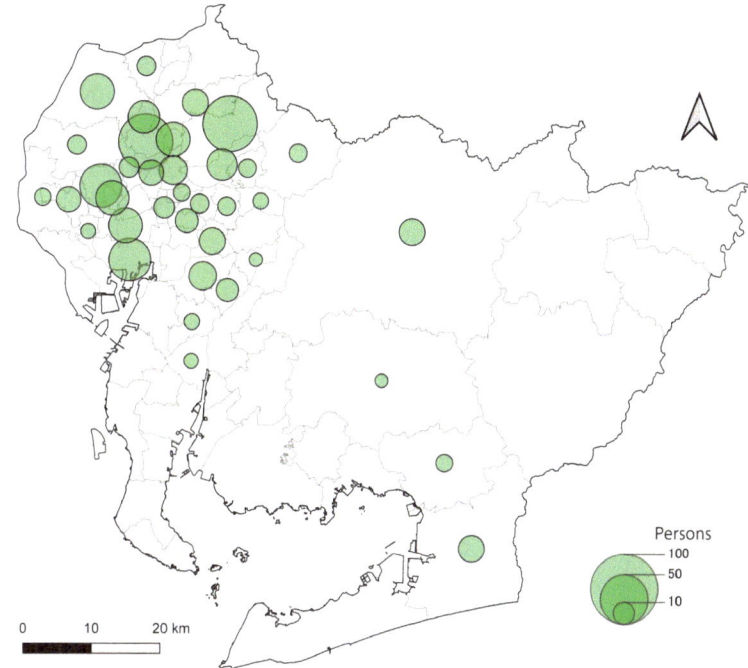

Fig. 1.3 Distribution of Turkish residents in Aichi Prefecture. *Note* Only populations of 5 or more people are shown. *Source* Microdata of the 2015 Population Census

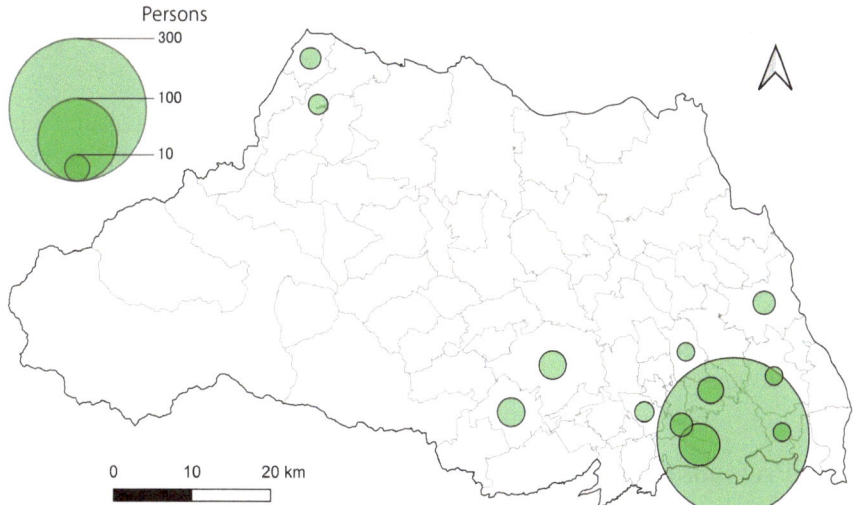

Fig. 1.4 Distribution of Turkish residents in Saitama Prefecture. *Note* Only populations of 5 or more people are shown. *Source* Microdata of the 2015 Population Census

1.3 Analytical Framework

According to Zelinsky and Lee (1998), heterolocalism has the following five characteristics:

(i) Spatial dispersion of residence by immigrants of a particular nationality

   Soon after entering the host country, immigrants choose to either spatially disperse or at most cluster to a modest degree, creating heterolocal conditions. This is made possible by their high economic status and access to advanced communication technology and means of transportation, although not all heterolocal groups belong to the middle or higher socioeconomic classes.

(ii) Spatial disjuncture between home and work

   Immigrants usually live and work in separate areas. Many full-time or part-time workers earn wages by commuting to work using the available means of transportation. Their places of residence are also often separated from where they shop and engage in social activities.

(iii) Formation of ethnic community without propinquity

   Once their personal networks become functional, they are able to create and maintain communities without the official activities as an organization. The emergence and rapid advances of the internet in the 1990s have led to the creation and maintenance of strong ethnic community ties on the metropolitan, regional, national, and even global scale, despite the absence of propinquity.

(iv) Emergence of heterolocalism as a time-dependent phenomenon

   Heterolocalism is a phenomenon that changes with time. Although earlier tendencies in the direction of heterolocalism were identified, its full-blown development only became possible under the socioeconomic and technological conditions existing at the end of the twentieth century.

(v) Formation of networks beyond the metropolis

   In regard to the formation of networks beyond the metropolitan spatial scale, Zelinsky and Lee argued that heterolocalism can be linked to transnationalism. This is said to be made possible by the innovations in information and communications technologies, such as the internet, which expand human action spaces, creating a transnational as well as a heterolocal environment.

   Zelinsky and Lee (1998), however, did not conduct any specific analysis using empirical data, nor did they examine any specific ethnic group. Although their model is premised on the American situation, we believe it is also applicable to other countries (Ishikawa 2019).

1.4 Previous Studies

The following papers are empirical studies using the heterolocal model to analyze immigrant residential patterns.

   Hardwick and Meacham (2005) demonstrated that residential dispersion occurred from the early stages of migration by new Vietnamese refugees into Portland, Oregon.

In other words, they found that the immigrants exhibited heterolocal residential patterns, which could no longer be explained by conventional assimilation models.

Hardwick (2006) mapped and analyzed the residential patterns of immigrants in the U.S. and Canada and found that their residential dispersion did not take place in the manner predicted by Zelinsky and Lee (1998), but rather through the formation of a large number of dispersed, small-scale settlements. They named this pattern "nodal heterolocalism.[3]"

Dennis (2007) verified the potential application of heterolocalism to Latino communities in the southeastern U.S. through quantitative and qualitative analysis. He pointed out that although many ethnic communities are socially unified, they are not spatially coherent, demonstrating that conventional theories that consider spatial proximity an essential component in creating and maintaining a unified ethnic society no longer apply in today's world where the distances in human interaction have been dramatically reduced by technology.

Using census data, Kimber (2010) examined the heterolocal model through a longitudinal study of immigrants in Syracuse, New York. She argued that heterolocalism was not applicable prior to the change in immigration laws in 1965 because immigrants tended to live in urban settlements by country or region of origin; however, she did find it applicable to immigrants after 1965 since they have usually moved to dispersed settlements immediately after their entry. Similarly, Halfacree (2012) suggested that new immigrant residential patterns cannot be analyzed under conventional theoretical frameworks, using the term "dynamic heterolocalism" to emphasize the dynamic characteristic of heterolocalism, i.e., building identities across multiple locations.

Bushi (2014) studied the communities of Albanian immigrant groups in Chicago and New York by discussing their diverse organizational roles. Through case studies on immigrants, she claimed that heterolocalism can be applied to some extent in a situation she termed "segmented heterolocalism," which takes into account the frequency of interactions within the community, the actors carrying out the interactions, and the persistence of the interactions.

In Japan, Sugiura (2011) and Fukumoto (2018) introduced the theory of heterolocalism, while Ishikawa proposed explanatory frameworks for ethnic enclaves, segregation, and related changes in urban areas. These frameworks included heterolocalism, spatial assimilation, and pluralism. In a 2019 paper, Ishikawa explained the importance of examining the details of the diverse reality surrounding the settlement of ethnic groups, as minorities in the host country on the basis of these frameworks and in accordance with concrete data. Ishikawa substantiated these claims in a book published in 2021. In the following, we discuss the Turkish residential patterns in Aichi Prefecture using the research method of Ishikawa (2021).

---

[3] The term "nodal heterolocalism" was clearly defined by Hardwick (2006), although it had been used by Hardwick and Meacham (2005).

1.5   Empirical Validity of Heterolocalism

(1)   Spatial dispersion of residence

Figure 1.3 shows the distribution of the Turkish residents in Aichi Prefecture. Spatial dispersion of residence is observed throughout the western part of Aichi Prefecture, indicating a "nodal heterolocal" situation with many dispersed small-scale settlements.

Ordu Province has 19 districts, but the immigrants in Aichi Prefecture are mostly from Fatsa, Ulubey, Kabadus, and Ordu Districts (Takeshita and Hanaoka 2015; Takeshita 2016). The first Turks who entered the labor market in Aichi Prefecture were the three pioneer migrants from Fatsa, Ordu Province, who came to the prefecture around 1990 with the purpose of finding employment (Milliyet Newspaper, December 19, 2003). A succession of relatives and friends later followed them to Japan with their help. Japan and Türkiye have a visa waiver agreement, which allows Turks to enter Japan for sightseeing purposes for up to 90 days without a visa. Consequently, Japan, especially Aichi Prefecture, became a new work destination among the people of Ordu Province (Takeshita 2021).

Although structural conditions, such as economic and institutional factors, have a significant impact on the decision to migrate as a worker from Türkiye to Japan, the theory of migration systems can explain why migration specifically took place from Ordu Province to Aichi Prefecture. Japan's migration systems include the market brokerage migration system and the mutual assistance migration system (Kajita et al. 2005). The market brokerage migration system is roughly divided into two types, namely "predecessors with migration experience receiving compensation for providing knowledge and knowhow" and "intervention by more professional and commercial brokers in the migration process." In the mutual assistance migration system, predecessors help their successors, who further help later successors to reciprocate the favor received from the predecessors. By using social networks, it is possible to reduce the costs and risks that accompany relocation; for example, in securing a place of residence and work at the new destination (Massey et al. 1993).

Many Turks in Ordu Province set their sights on Aichi Prefecture, either as a result of having taken seriously the exaggerated success stories from those who had returned home or having seen the fine houses the returnees had built in Ordu. Those who came early on, especially those who obtained residency status by marrying a Japanese citizen, started up personal side businesses as brokers, forming the base for the market brokerage migration system. On the Turkish side, the contact persons for migration brokerage are the fathers or brothers who remain in Ordu. The market brokerage migration system between Ordu and Aichi Prefecture was not the result of the intervention of a professional and commercial broker, so there was less entry through this system, which also led to the spatial dispersion of residence.

As more of their relatives came to Japan through the market brokerage migration system, the number of people from Ordu Province further increased also through the mutual assistance migration system that was dependent on those who were already in Japan. Chain migration is more likely to occur among people from agricultural regions than those from urban areas (Wilpert 1992). Therefore, it was easy to form a migration system where predecessors helped their successors among the people of Ordu who lived in close family and neighborhood networks.

Although the visa waiver agreement with Türkiye remains in effect, recently, entry procedures for visiting relatives and tourists have become more restrictive. Consequently, most Turks newly arriving from Ordu Province come to Japan through invitation as workers arranged by brothers or relatives who had earlier arrived from Ordu and established their own demolition business as described below. In other cases, relatives are invited as workers by earlier immigrants who had started their own Turkish restaurant or used car export business. Even though residentially dispersed Turks invite relatives to either live with them or live close by, spatial dispersion of residence is maintained due to the small number of immigrants.

(2)   Spatial disjuncture between home and work

As shown in Table 1.1, nearly half of the Turkish men in Aichi Prefecture are classified as construction and mining workers. This is because demolition workers are included in the category of construction and mining workers, and immigrant workers from Ordu Province form a niche of demolition industry in Aichi Prefecture. Most of them are high school graduates or lower in educational attainment and cannot speak Japanese at all when they come to Japan. While this occupation is more dangerous and unstable than working in a factory, the ability to easily land a job through the network of people from the same home province has resulted in the concentration of Turkish workers engaged in demolition work.

Japanese people have not been inclined to engage in demolition work, which led to a workforce shortage. Therefore, employers tended to hire foreigners illegally, even though they knew that they would be punished if caught by the Immigration Services Bureau. This was how Turks from Ordu Province have developed their own labor niche.

Table 1.1 Occupations of Turkish men

Occupation	Aichi Prefecture		Japan	
	Actual number (persons)	Percentage (%)	Actual number (persons)	Percentage (%)
Construction and mining workers	175	44.0	279	24.9
Manufacturing process workers	48	12.1	199	17.7
Transport and machine operation workers	28	7.0	43	3.8
Service workers	14	3.5	123	11.0
Professional and engineering workers	14	3.5	87	7.8
Others	119	29.9	391	34.8
Total	398	100.0	1122	100.0

Source Microdata of the 2015 Population Census

Although the model of Zelinsky and Lee (1998) does not clearly define the "spatial disjuncture" between residence and workplace, in this case they are in fact separated because demolition work has no fixed workplace, and thus work is not necessarily available near one's place of residence.

(3) Formation of ethnic community without propinquity

As the Turks began to stay longer in Japan, they expanded and strengthened their networks. Their shared networks formed a composite community based on a Muslim network, a network of immigrants from Ordu Province, and a network of demolition workers. They were also held together by the common challenges that came with their prolonged stay in Japan.

In the 2000s, as they continued to settle down through marriage with Japanese women,[4] they began to form their own community. As their life stages shifted, the challenges they faced changed from workers' issues to long-term residents' issues, including the problem of Islamic education for their children (Takeshita 2016). This led to the heightened importance of a community that could cater to needs that significantly differ from those of the host society. The community they formed, however, was not one composed of people living close together but a network-based community with dispersed members.

The core of their community is the Kakamigahara Mosque, which opened in 2004. Turks from Ordu Province revamped a karaoke establishment to open their own mosque where sermons are delivered in Turkish. Mosques in Japan are not necessarily located in Muslim enclaves (Okai 2009: 25). The Kakamigahara Mosque is located in Kakamigahara City, Gifu Prefecture, where real estate prices are relatively low. Kakamigahara is close to the northwestern part of Aichi Prefecture and about a 40-min drive from Kitanagoya City where many Turks live. This Mosque is used not only for religious practices but also as a place for the congregants to exchange information and help each other in everyday life.

Heterolocalism does not preclude the idea that formal and informal ethnic facilities are concentrated in certain areas within a city and represent important identification points for the ethnic community. However, since these facilities can be accessed through public or private transportation, even from more distant areas, it is likely that ethnic community structures are also dispersed beyond these localities (Farwick 2011: 248–249). Accordingly, the Kakamigahara Mosque can be considered the symbolic center of the Turkish ethnic community.

Likewise, the role of Turkish restaurants should not be overlooked. Although there are many Turkish restaurants in Aichi Prefecture that target Japanese customers, most of the Turks from Ordu Province come to the Turkish restaurants in northwestern Aichi Prefecture. Almost all of the customers of the two Turkish restaurants in Kitanagoya City and in Toyoyama Town are Turks from Ordu, and the Turkish restaurant in the southwestern part of Komaki City near Toyoyama Town also has many

---

[4] According to the microdata of the 2015 Population Census, in Aichi Prefecture, the proportion of Turkish women married to Japanese men was 11.3%, and that of Turkish men married to Japanese women was 67.1%, showing a high proportion of intermarriage between Turkish men and Japanese women.

customers from Ordu. These restaurants are run by Turks from Ordu. In particular, the Turkish restaurant in Toyoyama Town offers tables for playing *Okey*,[5] making it a place for rest and relaxation for many Turkish men from Ordu. Some drive over an hour just to get there. There is also a *baklava* (traditional Turkish pastry) shop in Aisai City, although this is unfamiliar fare in Japan. A Turk from Safranbolu opened the shop in Tokyo, but in 2019, the owner moved the shop to Aisai City due to the high demand for *baklava* among the many Turks living in Aichi Prefecture.[6]

The Turkish community in Aichi Prefecture was not formed through a clearly defined enclave. Despite the lack of spatial proximity, the community has maintained strong ties among members not only because of their common ethnic identity but also because the community was formed through a combination of factors, namely religion, place of origin, and type of occupation.

(4) Emergence of heterolocalism as a time-dependent phenomenon

As globalization progressed around the turn of the twentieth century, events, decisions, and activities in other parts of the world became relevant to individuals and communities in faraway places. Social, political, and economic activities started to transcend national borders. In this context, globalization has brought about interconnectedness across regions, the expansion of networks of social activity and power, and the possibility of engagement from remote locations (Held et al. 1999: 15).

Immigrants today generally maintain networks with relatives back home even as they establish new lives and try to survive economically in a new land, moving spatially to live their lives while building cross-border social relationships. Transnational migration has been propelled by advances in communication technologies and transportation, and the lives and livelihoods of migrants have been greatly enhanced by technological innovations (Ishikawa 2012: 25).

In particular, the rapidly spreading internet has had a significant impact on heterolocal residential patterns. The internet began to penetrate Japan in the latter half of the 1990s, about the same time as the rapid increase in the number of Turks migrating to Aichi Prefecture from Ordu Province. The internet became widespread throughout Japan in the 2000s, while people in Ordu had to visit a local internet café to use the internet. Later, smartphones became the main means of communication, making it easier for these immigrants to communicate via video calls and social media with their relatives back in Ordu, as well as with their fellow countrymen in Japan. That technological leap paved the way for the creation of a heterolocal conditions.

(5) Formation of networks beyond the metropolis

In this transnational age, the heterolocalism and transnationalism models help explain how recent immigrants balance living in two different cultural environments. They also help explain how immigrants creatively interweave the two different cultures to establish their own identities and lifestyles (Al-Huraibi 2009: 72).

---

[5] A traditional tile game popular in Türkiye that features 106 numbered tiles. Ideally, the game involves four players.

[6] Based on an interview held in March 2020 with the manager of the baklava specialty store.

The main differences between heterolocalism and transnationalism are as follows. In heterolocalism, ethnic groups are strongly connected by internal networks, despite the absence of ethnic enclaves (Zelinsky and Lee 1998: 289). Transnationalism, on the other hand, refers to the process by which migrants build and maintain multiple social relationships that connect their societies of origin to their host societies. Many immigrants create social spheres that transcend geographical, cultural, and political boundaries (Basch et al. 1994: 8).

Many studies of Turkish immigrants in Germany have explored the implications of transnationalism theory (Aksoy and Robins 2000; Çağlar 2001; Argun 2003; Østergaard-Nielsen 2003; Ehrkamp 2005; Ishikawa 2012). According to Ehrkamp, Turkish immigrants in German cities maintain close ties with their home country and/or the place where they used to live, leading to the transnationalization of spaces (Çağlar 2001; Ehrkamp 2005). Social processes and relations not only create a place in a material sense but also produce the meaning that people attach to places, evoking a sense of place (Massey 1994). Turkish mass media are widely available, and most Turkish residents maintain familial ties to Türkiye. Turkish communities reflect transnational practices as well as local attachments through immigrants' investments in the local neighborhood (Ehrkamp 2005).

Turkish immigrants have networks of relatives and friends in Aichi Prefecture similar to what they had in Ordu Province, while they have created a transnational network to maintain close ties with relatives and friends in Ordu. If a Turkish resident in Aichi Prefecture owns a home or manages an apartment building in Ordu, they would typically turn to their fathers or brothers in Ordu to collect the rent or supervise the construction. Managing an apartment building in Ordu from overseas is a form of investment in one's home country, and this too establishes a link between the immigrant community and the Turkish homeland. As mentioned above, the widespread availability of the internet has made it very easy to form and maintain transnational networks.

With the opening of Ordu-Giresun Airport in 2015, travel between Istanbul and Ordu was shortened to 1 h and 40 min, greatly reducing the time for traveling between Japan and Ordu via Istanbul. Many of the people who came to Japan in the 1990s had to take a 12-h bus ride from Ordu to Istanbul and another 12-h flight to Japan. At the time, the shortest way to travel between Ordu and Istanbul was to have a relative drive the migrant from Ordu to Samsun Airport in three hours and then fly from Samsun Airport to Istanbul Airport in an hour and a half. After that, with the opening of the expressway, the connection between Samsun and Ordu was shortened to an hour and a half by car. The new expressway has made it easier for Turks to return to their hometown, but the opening of Ordu-Giresun Airport has further shortened that distance.

1.6 Conclusion

The following is a summary of the insights gained from this study. After the three pioneer migrant workers from Ordu Province, Türkiye came to Aichi Prefecture in the 1990s, many young men have since migrated from Ordu to Aichi Prefecture with the aim of finding work. Many of them became engaged in demolition work, eventually forming a niche industry for demolition work in Aichi Prefecture. In the 2000s, they continued to settle down through marriage with Japanese women, and in the 2010s, many of them set up their own demolition companies. They then invited their relatives from Ordu to work for them. Since this migration from Ordu to Aichi Prefecture did not involve professional and commercial brokers, the number of immigrants was small, and this also led to spatially dispersed settlements. Since then, Turks from Ordu, who were scattered in the northwestern part of Aichi Prefecture, have invited their own relatives, resulting in many dispersed small-scale settlements, i.e., the creation of a "nodal heterolocal" conditions.

These immigrants' homes and workplaces are separated because demolition work has no fixed workplace, and thus work is not necessarily available near one's place of residence. They have been held together by a complex of shared networks: They share the same religion (Islam), they migrated from the same hometown (Ordu), and they engage in the same line of work (demolition). Consequently, they have formed a network-based community made up of members living in dispersed residences. In other words, they created and maintained an ethnic community without propinquity, while creating and maintaining a transnational network with relatives in their hometown. These conditions were made possible by advanced communication technologies; the widespread use of the internet in particular has had a significant impact on the development of heterolocalism.

Therefore, the heterolocal model of Zelinsky and Lee provides a good explanation of the residential patterns of Turks in Aichi Prefecture. Networks of migrants who share religious and ethnic identities are creating heterolocal conditions on a variety of scales across regions and borders.

Going forward, we will continue to study how the residential patterns of Turks in Aichi Prefecture change over time, and whether the patterns of the second generation of Turks will be different from those of the first generation.

Acknowledgements The authors would like to express their sincere gratitude to the National Statistics Center for providing the microdata on foreigners from the 2015 Population Census.

References[7]

Abe R (2011) Esunishitei no chirigaku: imin esunikku kukan wo tou (Geography of ethnicity: exploring ethnic space of immigrants). Kokon Shoin, Tokyo (J)

Aksoy A, Robins K (2000) Thinking across space: transnational television from Turkey. European Journal of Cultural Studies 3(3): 343–365. https://doi.org/10.1177/136754940000300305

Al-Huraibi N (2009) Islam, gender, and integration in transnational/heterolocal contexts: a case study of Somali immigrant families in Columbus, Ohio. Dissertation, Ohio State University. https://www.academia.edu/1958328/

Argun BE (2003) Turkey in Germany: the transnational sphere of Deutschkei. Routledge, London

Basch L, Schiller NG, Blanc CS (1994) Nations unbound: transnational projects, postcolonial predicaments, and deterritorialized nation-states. Routledge, London

Bushi M (2014) Rethinking heterolocalism: the case of place-making among Albanian-Americans. Geography Honors Projects, Paper 40. https://core.ac.uk/download/pdf/46725534.pdf

Çağlar AS (2001) Constraining metaphors and the transnationalism of space in Berlin. Journal of Ethnic and Migration Studies 27(4): 601–613. https://doi.org/10.1080/13691830120090403

Dennis K (2007) Testing heterolocalism: an assessment of Latino settlement patterns in the Southern United States. Master's Thesis, University of Tennessee. https://trace.tennessee.edu/cgi/viewco ntent.cgi?article=1147&context=utk_gradthes

Ehrkamp P (2005) Placing identities: transnational practices and local attachments of Turkish immigrants in Germany. Journal of Ethnic and Migration Studies 31(2): 345–364. https://doi.org/10. 1080/1369183042000339963

Farwick A (2011) The effect of ethnic segregation on the process of assimilation. In: Matthias W, Windzio M, de Valk H, Aybek C (eds) A life-course perspective on migration and integration. Springer, New York, p 239–258. https://doi.org/10.1007/978-94-007-1545-5

Fukumoto T (2013) The persistence of the residential concentration of Koreans in Osaka from 1950 to 1980: its relation to land transfers and home-work relationships. Jimbun Chiri (Japanese Journal of Human Geography). 65(6): 475–493. https://doi.org/10.4200/jjhg.65.6_475

Fukumoto T (2018) Esunikku seguregeshon kenkyu ni kansuru oboegaki (A brief note on ethnic segregation studies: implications for empirical research on Japan). Kukan, Shakai, Chiri-shiso (Space, Society and Geographical Thought) 21: 15–27 (J)

Halfacree K (2012) Heterolocal identities? Counter-urbanisation, second homes, and rural consumption in the era of mobilities. Population, Space and Place 18: 209–224. https://doi.org/10.1002/ psp.665

Hardwick SW (2006) Nodal heterolocalism and transnationalism at the United States-Canadian Border. The Geographical Review 96(2): 212–228. https://doi.org/10.1111/j.1931-0846.2006. tb00050.x

Hardwick SW, Meacham JE (2005) Heterolocalism, networks of ethnicity, and refugee communities in the Pacific Northwest: the Portland story. The Professional Geographer 57(4): 539–557. https://doi.org/10.1111/j.1467-9272.2005.00498.x

Held D, McGrew A, Glodblatt D, Perraton J (1999) Global transformation: politics, economics and culture. Palgrave Macmillan, London

Immigration Services Agency of Japan (2020a) Reiwa gan-nen matsu ni okeru zairyu gaikokujinsu ni tsuite (The number of foreign residents in 2019). http://www.moj.go.jp/nyuukokukanri/kou hou/nyuukokukanri04_00003.html. Accessed 29 Feb 2020 (J)

Immigration Services Agency of Japan (2020b) Reiwa gan-nen ban zairyu gaikokujin tokei (Statistics on foreigners registered in Japan). https://www.e-stat.go.jp/stat-search/files?page= 1&layout=datalist&toukei=00250012&tstat=000001018034&cycle=1&year=20190&month= 24101212&tclass1=000001060399. Accessed 7 Aug 2020 (J)

---

[7] (J): Written in Japanese

Ishikawa S (2012) Doitsu zaiju toruko imin no bunka to chiiki-shakai: shakaiteki togo ni kansuru bunkajinruigaku teki kenkyu (Regional society and culture of Turkish immigrants in Germany: an anthropological study of social integration). Yuhikaku, Tokyo (J)

Ishikawa Y (2005) Gaikokujin kankei no ni-tokei no hikaku (Comparing the number of foreign residents between the two major population statistics in Japan). Jinkogaku Kenkyu (The Journal of Population Studies) 37: 83–94. https://doi.org/10.24454/jps.37.0_83 (J)

Ishikawa, Y (2019) Esunikku shudan no toshinai shujuchi ni kansuru kenkyu doko: beikoku deno seika wo chushin ni (Review of existing literature on ethnic enclaves: focusing on results obtained in the US). Ritsumeikan Chirigaku (The Journal of Ritsumeikan Geographical Society) 31: 1–12 (J)

Ishikawa Y (ed) (2021) Ethnic enclaves in contemporary Japan. Springer, Singapore

Ishikawa Y, Hanaoka K (2021) Overview of ethnic enclaves as example cases. In: Ishikawa Y (ed) Ethnic enclaves in contemporary Japan. Springer, Singapore, p 17–44. https://doi.org/10.1007/978-981-33-6995-5_2

Kajita T, Tanno K, Higuchi N (2005) Kao no mienai teijuka: nikkei burajirujin no kokka, shijo, imin nettowaku (Invisible residents: Japanese Brazilians vis-à-vis the state, the market and the immigrant network). Nagoya University Press, Nagoya (J)

Kataoka H (2013) Concentrated ethnic towns and dispersed/assimilated ethnic towns: regional disparities in the formation and development of ethnic towns: case studies of Brazilian residents in Japan. Jimbun Chiri (Japanese Journal of Human Geography) 65(6): 494–507.https://doi.org/10.4200/jjhg.65.6_494

Kimber L (2010) New immigrant settlement patterns in Syracuse, NY: an assessment of the model of heterolocalism. Upstate Institute Student Research, Paper 7. http://blogs.colgate.edu/upstateinstitute/files/2013/01/Kimber-heterolocalism.pdf

Massey D (1994) Space, place, and gender. University of Minnesota Press, Minneapolis

Massey DS., Arango J, Hugo G, Kouaouci A, Pellegrino A, Taylor JE (1993) Theories of international migration: a review and appraisal. Population and Development Review 19(3): 431–466.https://doi.org/10.2307/2938462

Okai, H (2009) Tainichi musurimu niyoru shukyoteki kiban no kakutoku to henyo: mosuku setsuritsu katsudo wo chushin ni (Establishment of religious institutions by Muslim immigrants in Japan). Ningen Kagaku Kenkyu (Waseda Journal of Human Sciences) 22: 15–29 (J)

Østergaard-Nielsen EK (2003) Transnational politics: the case of Turks and Kurds in Germany. Routledge, London

Sugiura T (2011) Imin shudan no segurigeshon to esunisitei henyo (Segregation and changing ethnicity of immigrant groups). In: Yamashita K (ed) Gendai no esunikku shakai wo saguru: riron kara firudo he (Exploring contemporary ethnic community: theory and fieldwork) Gakubun Sha, Tokyo, p 39–46 (J)

Takahata S (2012) Daitoshi no hankagai to imin josei: Nagoya-shi Naka-ku Sakae chiku no firipin komyunitei ha nani wo kaetaka (Changes brought about by Filipino women in the Sakae-higashi area of Nagoya city). Shakaigaku Hyoron (Japanese Sociological Review) 62(4): 504–520. https://doi.org/10.4057/jsr.62.504 (J)

Takeshita S (2016) Social and human capital among Japanese-Turkish families in Japan. Asian Ethnicity 17(3): 456–466. https://doi.org/10.1080/14631369.2015.1062071

Takeshita S (2021) Turkish residents and marital assimilation. In: Ishikawa Y (ed) Ethnic enclaves in contemporary Japan. Springer, Singapore, p 153–177. https://doi.org/10.1007/978-981-33-6995-5_7

Takeshita S, Hanaoka K (2015) Turkish families and Islamic education for children in Aichi prefecture. In: Ishikawa Y (ed) International migrants in Japan: contributions in an era of population decline. Trans Pacific Press, Melbourne, p 195–211

Wilpert C (1992) The use of social networks in Turkish migration to Germany. In: Kritz MM, Lim LL, Zlotnik H (eds) International migration systems: a global approach. Oxford University Press, London, p 177–189

Wright R, Ellis M (2000) Race, region and the territorial politics of immigration in the US. International Journal of Population Geography 6: 197–211.https://doi.org/10.1002/1099-1220(200 005/06)6:3%3C197::aid-ijpg183%3E3.0.co;2-f

Yamashita K (2019) Sekai no chaina taun no keisei to henyo: firudowaku kara kajin shakai wo tankyu suru (Formation and change of the world's Chinatowns: exploring Chinese society through fieldwork). Akashi Shoten, Tokyo (J)

Yoshida M (2011) Women, citizenship and migration: the resettlement of Vietnamese Refugees in Australia and Japan. Nakanishiya Shuppan, Kyoto

Zelinsky W, Lee BA (1998) Heterolocalism: an alternative model of the sociospatial behaviour of immigrant ethnic communities. International Journal of Population Geography 4(4): 281–298. https://doi.org/10.1002/(sici)1099-1220(199812)4:4%3C281::aid-ijpg108%3E3.0.co;2-o

Chapter 2
Examining Geographic Distribution of LGBTs in Osaka City, Japan

Masakazu Yamauchi

Abstract Considerable research has been conducted in Western countries on the geographical patterns of sexual and gender minority populations (gay, lesbian, bisexual, and transgender populations, abbreviated as "LGBTs") and the enclaves they have created in large cities. However, the geographical distribution of LGBTs in Japanese cities has not been studied. This paper presents a quantitative study of whether LGBTs are unevenly distributed geographically in Osaka City, the central city of the second-largest metropolitan area in Japan. The study uses microdata from a 2019 random sampling survey on sexual and gender minorities in Osaka City, titled the "Survey on Diversity of Work and Life, and Coexistence among the Residents of Osaka City," which is one of the first representative random sampling surveys with questions on sexual orientation and gender identity. It was found that LGBTs are distributed unevenly in relation to non-LGBTs. However, when we controlled respondents' demographic and socioeconomic variables, including age, gender, education, number of household members, occupation, and years of residence in Osaka City, logistic regression models showed no significant association between the presence of LGBTs among respondents and the distinct areas where LGBTs were found to be concentrated. Therefore, we concluded that an uneven distribution of LGBTs in Osaka City may not actually reflect the geographical context of LGBTs concentration but rather mirror general variations in population composition.

Keywords Sexual and gender minority · LGBT · Geographic pattern · Osaka City

2.1 Introduction

In recent years, there has been a growing interest in investigating the scale of sexual and gender minority populations (gay, lesbian, bisexual, and transgender populations, abbreviated as "LGBTs") as well as their socioeconomic experiences across

M. Yamauchi (✉)
Faculty of Education and Integrated Arts and Sciences, Waseda University, Shinju-ku, Tokyo, Japan
e-mail: yamauchi-masa@waseda.jp

Y. Ishikawa (ed.), *Japanese Population Geographies II*,
Population Studies of Japan, https://doi.org/10.1007/978-981-99-2076-1_2

countries through quantitative approaches (Hiramori and Kamano 2020). Due to the limited availability of data collected by population-based surveys that ask questions on sexual orientation and gender identity, quantitative research on LGBTs has relied mainly on non-representative data. However, representative surveys to explore the proportion of LGBTs within the general population and the socioeconomic situations of LGBTs in comparison with the non-LGBT population have begun to be conducted in Western countries. Accordingly, scholars have pursued topics germane to LGBTs with statistical comparisons between LGBTs and non-LGBTs (e.g., Baumle 2013). The geographic distribution of LGBTs, which is one such topic, has been studied in Western countries.

On the national scale, these studies have shown that gay and lesbian populations tend to concentrate in large cities (Black et al. 2000; Anderson et al. 2006; Duncan and Smith 2006; Gates 2013; Wimark and Östh 2014). For example, Black et al. (2000), using the 1990 census' public microdata samples, demonstrated that the top 20 cities in the United States with the highest number of *male same-sex couples* accounted for 59% of such households, while these cities accounted for only 26% of the United States' population; meanwhile, they showed that the top 20 cities in the United States with the highest number of *female same-sex couples* accounted for 45% of such households, and these cities accounted for only 25% of the United States' population. Wimark and Östh (2014) examined the geographic distribution of gay and lesbian singles and couples in Sweden using the information from internet sites that have large memberships of lesbian, gay, and bisexual individuals. Their results indicated that in addition to gay and lesbian populations tending to concentrate in large cities, their distribution varied in accordance with the size of the population of the region, and singles were more likely to concentrate in large cities than couples.

On the urban scale, studies have shown that gay and lesbian populations tend to concentrate in specific neighborhoods within cities (Anacker and Morrow-Jones, 2005; Goldie 2018). For example, Goldie (2018) demonstrated using census data that male and female same-sex couples in Melbourne and Sydney, Australia, were concentrated in distinct central neighborhoods that are segregated from each other but geographically close. Detailed studies have also been conducted on distinct neighborhoods where gay and lesbian populations were concentrated and built their communities (Schroeder 2014; Kanai and Kenttamaa-Squires 2015; Smart and Whittemore 2017). For example, while Smart and Whittemore (2017) used real estate listings from the gay- and lesbian-oriented newspapers to show that the gay- and lesbian-oriented real estate market continued to focus on traditional gay and lesbian enclaves in central Dallas, Texas, Kanai and Kenttamaa-Squires (2015) utilized multiple types of materials to reveal that the characteristics of South Beach as a gayborhood changed to an LGBT-friendly mixed neighborhood increasingly under policies promoting tourism and redevelopment by the City of Miami Beach, Florida.

Blank and Rosen-Zvi (2012) argued three main factors for the high proportion of gays and lesbians tending to concentrate in large cities and in a limited number of enclaves within these cities. First, they desire to build communities of people with whom they can express and share their identity and culture without fear. Second, they feel the need for various amenities of living comfortably, such as restaurants and cultural facilities, and for a clean and healthy environment. Third, they want to participate in their community's legal system in order to gain representation in local government councils and other legal bodies as a way to strengthen their political influence and make it easier for them to achieve their desired lifestyle.

In contrast to these research findings for Western countries, the tendency of gay and lesbian populations to concentrate in distinct neighborhoods within large cities seems to be less common in non-Western countries. Describing the formation of a spatial concentration of gay venues such as bars, discos, and saunas in Singapore and Hong Kong, where homosexuality is not legally recognized while traditional Confucian ideas emphasizing heterosexuality and family are deeply rooted, Yue and Leung (2017: 758) argued, "There is thus no comparable formation of the kind of 'gay village' found in North America and Western Europe."

In Japan, just as in Singapore and Hong Kong, gay populations do not seem to build the kind of 'gay village' within large cities where there exists neighborhoods with a concentration of gay venues. For example, the Shinjuku Ni-chōme district in Tokyo is known to be a large gay consumer space that provides lifestyle options for LGBTs (Sunagawa 2015; Mitsuhashi 2018; Suzaki 2019a, b),[1] but most of its residents are not LGBTs (Sunagawa 2015). Kamiya (2018: 78) has also argued, "there are no neighborhoods where homosexual populations concentrate to reside in Japanese cities."

However, no empirical studies on the geographic distribution of LGBTs have been conducted in Japan. This is due in large part to the absence of an urban residential landscape of LGBTs similar to those in Western countries and to the dearth of social surveys that include questions on sexual orientation and gender identity that could enable discussions on their geographic distribution quantitatively. Although quantitative studies of LGBTs using social surveys are gradually increasing in Japan (Kamano 2018), explicit quantitative focus on the geographic aspects of LGBT lives has been limited. Thus, quantitative treatment of the geographic characteristics of LGBTs in Japan remains a key challenge.

Therefore, this chapter addresses the geographic distribution of LGBTs in Osaka City, the central city of the second-largest metropolitan area in Japan. Here, our analysis is guided by two fundamental research questions: Is there a tendency for LGBTs to concentrate in particular wards of Osaka City? Are there any geographical factors peculiar to the geographic distribution of LGBTs? Parallel to this, the geographic distribution of foreign residents, another distinct minority within the population, is concurrently examined as a way to confirm the plausibility of analyzing the small data samples of LGBTs. While this chapter uses random sampling survey data as introduced in Section 2.2, samples from LGBTs in these data are quite few, so the results of quantitative analysis are probably more difficult to interpret adequately. Foreign residents can be regarded as a small population group like LGBTs, and as shown below, sufficient statistical data and investigations are available on foreigners (e.g., Fukumoto 2010, 2018a). In this chapter, we intend to present insights into the

---

[1] A variety of eating and drinking establishments can be found in Shinjuku Ni-chōme, which is also a gathering place for non-gay sexual minorities and heterosexuals (Sunagawa, 2015).

geographic distribution of LGBTs in Japan, which has not been studied thus far, as well as to demonstrate the practicability of investigating the geographic distribution of small population groups using data from a random sampling survey.

This chapter is organized as follows. Section 2.2 describes the data and methods used, and Sect. 2.3 shows the results of our data analysis. Section 2.4 discusses the validity of the analysis method used in this study and the geographic distribution of LGBTs in Osaka City. Section 2.5 summarizes our results.

2.2 Data and Methods

1. Data

The data used in this study are microdata from the "Survey on Diversity of Work and Life, and Coexistence among the Residents of Osaka City" (hereinafter, the Osaka City Residents' Survey, or OCRS) conducted in January 2019 with the cooperation of the Osaka City government (Kamano et al. 2019) by the research team of the project "Demography of Sexual Orientation and Gender Identity: Building a Foundation for Research in Japan," which includes the author (Principal Investigator: Saori Kamano). There are two reasons for selecting Osaka City as the target of the survey. One is that a large number of LGBTs are assumed to reside there, as the central city of the second-largest metropolitan area in Japan. Another is that the research teams received cooperation in the implementation of the survey from the Osaka City government, which has made efforts to eliminate prejudice and discrimination against minorities such as LGBTs.

The Osaka City Residents' Survey was one of first randomized social surveys to ask about sexual orientation and gender identity in Japan. In addition to sexual orientation and gender identity, various attributes, attitudes, past experiences, etc. of the respondents and their household members, as well as geographic information based on their current residential wards, were gathered by the survey. The questionnaire was sent to 15,000 people (about 1% of the residents), aged 18 to 59, who were randomly selected from the Basic Resident Registration (BRR) of Osaka City as of October 1, 2018.

The survey was conducted using a mixed-mode method in which a set of survey-related materials was mailed to the respondents, who were then able to choose whether to send their responses by mail or via the web using a unique ID and password. Furthermore, in order to elicit cooperation from foreign respondents, the survey was implemented with survey materials and a website for submitting answers written in six foreign languages in addition to Japanese. The materials were delivered to a total of 14,838 respondents, excluding the 162 respondents for which the materials were returned as undeliverable. From these, valid responses were received from

4285 residents. The valid response rate, therefore, excluding the returned portion, was 28.9%.

2. Methods

Three types of analysis were conducted in this study. In the first analysis, sampling weights designed to correct for representativeness in the results of OCRS were built, and the usefulness of weighted samples was examined by comparing the results using unweighted and weighted samples.

The sampling weights were built to match the population distribution (sex, age, and ward) in the BRR (as of October 1, 2018), which is the survey population, using the responses to questions about sex at birth, age at time of the survey, and ward where the resident resides, which are included in the OCRS survey items. Samples who did not answer the required items were excluded in building sampling weights, and age was classified into five-year age groups. A total of 4128 samples were analyzed, which amounted to 96.3% of the 4285 valid response samples.

The results of unweighted and weighted samples were compared with the results of the most recent Population Census of Japan, which is considered nearly equivalent to the survey population. The nine different benchmarks compared were age, sex, marital status, nationality, household size, educational attainment, employment status, occupation, and the ward in which they reside. The results for those aged 20–59 in the 2015 Population Census were used for all benchmarks, except for educational attainment, for which results from the 2010 Population Census were used.[2]

The categorical distribution of responses and the dissimilarity index by benchmark were used for the comparison. The dissimilarity index value, on a scale of 0 to 100, refers to the magnitude of the gap between the OCRS and the Population Census; while a value of 0 means that there is no gap between them, a value of 100 means that they completely diverge. In other words, the smaller the gap between them, the closer the value is to 0. However, the gaps in the dissimilarity index values among different benchmarks should be interpreted with caution, since the value of the dissimilarity index itself is also affected by the difference in the number of categories for each benchmark.

In addition to the above, nationality was also compared between the foreign population distributions by ward in the OCRS (unweighted and weighed samples) and in the survey population from the BRR (as of October 1, 2018). This investigation was conducted because foreign residents can be regarded as a small population group like LGBTs; therefore, they can be considered a reference group for assessing the geographic distribution of LGBTs. For the comparison, the distribution map by ward and dissimilarity index was used to examine the usefulness of weighted samples.

In the second and third analyses, we examined the geographic distribution of LGBTs in Osaka City. As Fukumoto (2018b) pointed out, there are two approaches to examining the geographic distribution of a particular population: whether a single population is geographically concentrated and whether there is unevenness in the

---

[2] Comparison with educational attainment was done only for the 2010 Population Census, since educational attainment was not surveyed in the 2015 Population Census.

geographic distribution between two populations. The latter approach was adopted in this study because there were small samples of LGBTs in the OCRS, and it was difficult to directly analyze the geographic concentration of a single population group.

In the second analysis, we examined how the geographic distribution of LGBTs differed from that of non-LGBTs among the wards in Osaka City. In accordance with the procedure taken in Kamano et al. (2019), gay, lesbian, bisexual, and transgender samples were identified and classified together as LGBT, while the remaining samples were classified as non-LGBT. Individual samples having various sexual orientations and gender identities were grouped together into the single category of LGBTs because they were small in number. Although this decision was unavoidable, it nevertheless seems fairly reasonable given the fact that studies in Western countries have shown that gay and lesbian enclaves are not widely segregated (Goldie 2018) and that venues such as bars for non-gay sexual minorities also exist in the neighborhoods with a concentration of gay venues in Japan (Sunagawa 2015).

The sexual and gender minority percent ratio (SGM-PR) shown in the following equation was then determined for each ward:

$$\mathrm{PR}_w = p_w(1 - q_w)/q_w(1 - p_w),$$

where

PR_w SGM-PR for the ward,
p_w ratio of the ward's LGBTs to Osaka City's LGBTs, and
q_w ratio of the ward's non-LGBTs to Osaka City's non-LGBTs.

The value of SGM-PR is equal to 1 if the ratio of LGBTs in a given ward to the total population of Osaka City is equal to the ratio of LGBTs to the population of that ward, less than 1 if the latter is larger than the former, and above 1 if the former is larger than the latter. When the value of SGM-PR is 1.5 or more, it is considered indicative of a tendency for LGBTs to concentrate in that ward.

Moreover, as reference, the validity of the results of the OCRS was examined by determining a similar index (referred to as F-PR) regarding whether foreign residents tend to concentrate in particular wards based on the nationality information in the OCRS and comparing it with the results of a similar analysis using the population by nationality based on the BRR (as of October 1, 2018).

In the third analysis, in order to determine whether LGBTs tend to concentrate in particular wards, logistic regression analysis was used to estimate the odds ratios (ORs) and 95% confidence intervals (CIs) of independent variables comprised of explanatory and control variables for LGBTs. The dependent variable was defined as "1" if the respondent was LGBT and as "0" if the respondent was non-LGBT. The explanatory variable for a geographical factor was defined as "1" if the ward where the respondent resides was considered an area in which LGBTs were concentrated—3 wards of which the SGM-PR value was 1.5 or more—and "0" if the ward where the respondent resides was considered an area in which LGBTs were not concentrated—21 wards of which the SGM-PR value was less than 1.5. Control variables for demographic and socioeconomic status were as follows: sex at birth (male and female), age (20–29 years, 30–39 years, 40–49 years, and 50–59 years), size of household (one person and two or more persons), educational attainment (low: junior high school/high school [12 years or less of education], middle: junior [technical] college/vocational school [13–15 years], and high: university/graduate school [16 years or more]), occupation (white-collar, gray-collar, blue-collar, and not working), and years of residence (10 years or less and 11 years or more).

The estimated ORs were used to assess independent associations between the dependent and independent variables. We focused especially on the OR of the explanatory variable, since if there were a specific mechanism by which LGBTs are concentrated in certain wards, a significant independent association between dependent and explanatory variables would be apparent. Therefore, when the OR of the explanatory variable is statistically significant, a specific mechanism for the geographic distribution of LGBTs exists. In contrast, when the explanatory variable has an OR with marginal significance, the geographic distribution of non-LGBTs with demographic and socioeconomic attributes similar to those of LGBTs could be considered similar to the geographic distribution of LGBTs: this can be interpreted as the geographic distribution of LGBTs having no specific mechanism.

The estimations of ORs and CIs of logistic regression models were computed using the penalized maximum likelihood method (Firth 1993; Heinze and Schemper 2002), along with the logistf package of the statistical software R, version 3.6.1, to avoid a complete or quasi-complete separation state in which the estimator is not uniquely defined, which occurs when using the conventional maximum likelihood estimation. A total of 3992 samples (93.2% of the total valid response samples without non-response items) were used for the logistic regression analysis.

2.3 Results

1. Comparison of unweighted and weighed sample statistics

Table 2.1 shows the unweighted and weighed sample statistics for the OCRS and the Population Census statistics, and Table 2.2 shows the results of dissimilarity indices. For age, sex at birth, and ward, which were used in building the sampling weights, the weighted sample statistics were more similar to the Population Census statistics than were the unweighted ones. The weighted sample statistics and the Population Census statistics did not match perfectly because the latter pertained to information on the de facto population in 2015, whereas the former pertained to information on the *da jure* population in 2019.

For the other items, statistics for marital status and nationality for the weighted samples were closer to those of the Population Census than those for the unweighted ones, and the gap was also smaller. For example, the dissimilarity index for marital status of weighted samples was lower than that of unweighted ones. However, there were items for which the dissimilarity indices were relatively large and the weighted

Table 2.1 Basic characteristics of respondents to OCRS and Population Census of Japan

		OCRS			Population census
		Sample size	Non-weighted	Weighted	census
		n	%	%	%
Age	20–24	240	5.9	11.1	9.9
	25–29	371	9.1	12.3	12.3
	30–34	471	11.5	12.7	13.1
	35–39	550	13.5	12.6	13.4
	40–44	576	14.1	13.9	15.4
	45–49	653	16.0	15.0	13.7
	50–54	614	15.1	12.6	12.0
	55–59	603	14.8	9.7	10.1
	Total	4078	100.0	100.0	100.0
Sex at birth	Male	1681	41.3	50.1	49.6
	Female	2390	58.7	49.9	50.4
	Total	4071	100.0	100.0	100.0
Marital status	Never married	1283	31.6	37.7	39.3
	Married	2405	59.3	54.6	54.2
	Divorced	345	8.5	7.2	5.8
	Widowed	25	0.6	0.5	0.7
	Total	4058	100.0	100.0	100.0
Nationality	Japan	3952	97.1	96.5	96.3

		OCRS			Population census
		Sample size	Non-weighted	Weighted	census
		n	%	%	%
Occupation	White-collar	2013	59.5	56.8	45.8
	Grey-collar	995	29.4	31.0	32.8
	Blue-collar	377	11.1	12.2	21.4
	Total	3385	100.0	100.0	100.0
Ward	Kita	227	5.7	5.4	5.3
	Miyakojima	160	4.0	3.9	3.9
	Fukushima	126	3.1	3.1	3.0
	Konohana	96	2.4	2.4	2.4
	Chuo	194	4.8	4.5	4.4
	Nishi	182	4.5	4.2	4.1
	Minato	114	2.8	2.9	2.9
	Taisho	68	1.7	2.2	2.2
	Tennoji	140	3.5	3.0	3.0
	Naniwa	81	2.0	3.1	3.1
	Nishiyodogawa	113	2.8	3.6	3.5
	Yodogawa	299	7.5	7.0	7.0
	Higashiyodogawa	249	6.2	6.3	6.7
	Higashinari	130	3.2	3.2	3.0

(continued)

Table 2.1 (continued)

		OCRS			Population census
		Sample size n	Non-weighted %	Weighted %	%
	Foreign	120	2.9	3.5	3.7
	Total	4072	100.0	100.0	100.0
Size of household	1	808	19.8	19.3	26.3
	2 or more	3264	80.2	80.7	73.7
	Total	4072	100.0	100.0	100.0
Education	Low	863	28.4	27.7	50.1
	Middle	895	29.4	27.8	21.6
	High	1284	42.2	44.5	28.4
	Total	3042	100.0	100.0	100.0
Labor force	Working	3407	85.4	85.4	78.5
	Not working	584	14.6	14.6	21.5
	Total	3991	100.0	100.0	100.0

	OCRS			Population census
	Sample size n	Non-weighted %	Weighted %	%
Ikuno	160	4.0	4.4	4.4
Asahi	104	2.6	3.1	3.2
Joto	277	6.9	6.0	6.0
Tsurumi	195	4.9	4.0	4.1
Abeno	183	4.6	3.8	3.9
Suminoe	152	3.8	4.2	4.2
Sumiyoshi	225	5.6	5.3	5.5
Higashisumiyoshi	192	4.8	4.5	4.3
Hirano	254	6.3	6.7	6.8
Nishinari	88	2.2	3.3	3.2
Total	4009	100.0	100.0	100.0

Source OCRS, Population Census of Japan

Table 2.2 Dissimilarity indices for measurement gaps between OCRS and Population Census of Japan

	Non-weighted	Weighted
Age	10.1	3.1
Sex at birth	8.3	0.5
Marital status	7.8	1.8
Nationality	0.8	0.3
Size of household	6.4	7.0
Education	21.7	22.3
Labor force	6.9	6.9
Occupation	13.7	11.0
Ward	5.6	1.0

Source OCRS, Population Census of Japan

sample statistics were not closer to those of the Population Census: size of household, educational attainment, employment status, and occupation.

Figure 2.1 shows the map of population distribution by ward using unweighted and weighed samples regarding nationality for the OCRS and for the BRR as of October 1, 2018. For Japanese nationals, although the gaps were very small, statistics for the weighted samples were closer to the BRR compared with the unweighted samples. The dissimilarity index was 4.7 for the unweighted samples and 1.5 for the weighted ones. For foreign nationals, on the other hand, there were large gaps with the BRR for both the unweighted and weighted samples. However, the weighted samples appeared to be closer to those observed in the Basic Resident Registration. The dissimilarity index for the weighted samples was 23.5 compared with 24.4 for the unweighted ones.

2. Distribution of LGBTs by ward

Table 2.3 shows the SGM-PR and F-PR by ward. The SGM-PR values vary from the highest of 3.82 for Naniwa Ward followed by 2.33 for Kita Ward, 1.67 for Nishi Ward, 1.49 for Ikuno Ward, 1.34 for Konohana Ward, etc., to the lowest of 0.27 for Taisho Ward, with 0.28 for Suminoe Ward, 0.32 for Tennoji and Higashiyodogawa Wards, 0.44 for Sumiyoshi Ward, etc.

As for F-PR, there were gaps between the OCRS and the BRR. In some cases, such as Chuo Ward and Tennoji Ward, values were below 1.0 for the OCRS, in contrast to being above 1.0 for the BRR; this was vice versa in other cases, such as Miyakojima Ward and Yodogawa Ward. However, if the wards were divided into two groups as those with high and those with low F-PR, the OCRS and the BRR showed generally consistent results. Thus, when the wards were divided into those with F-PR values equal to or higher than the standard value of 1.5 used in the analysis in the next section and those with values lower than this value, the results for the OCRS

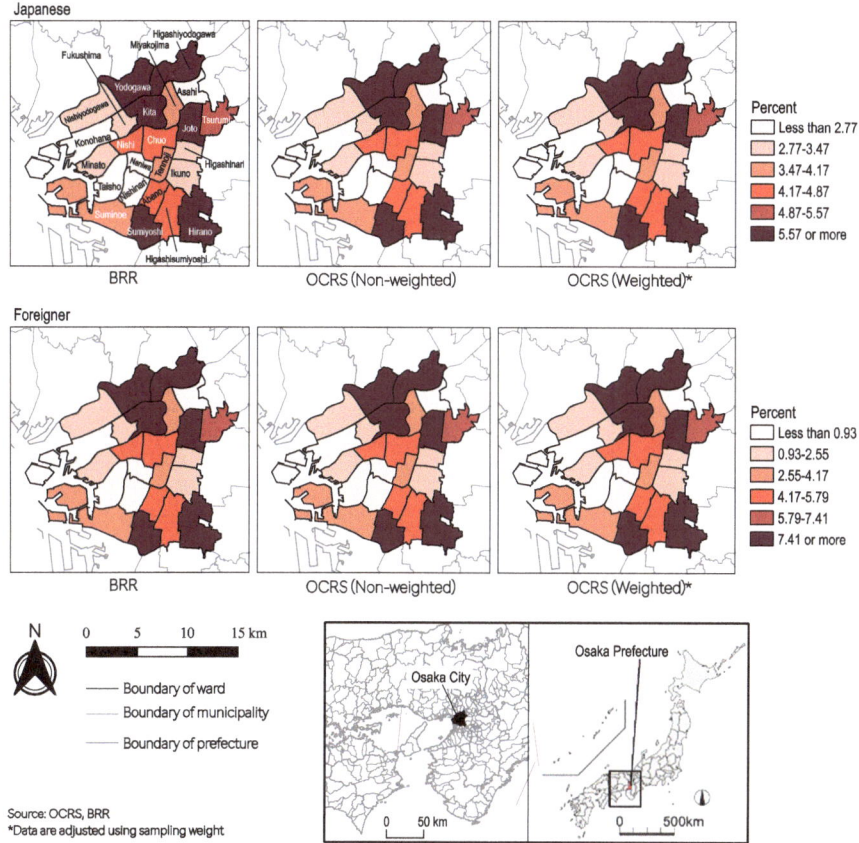

Fig. 2.1 Population distributions by ward in Osaka City

and the BRR matched for 21 wards but were not matched for 3 wards: Chuo Ward, Yodogawa Ward, and Higashinari Ward.

3. Results of logistic regression analysis

Table 2.4 illustrates the basic characteristics of the respondents and distribution of LGBTs by attributes. The proportion of LGBTs was 3.2% overall, 7.2% in wards with high SGM-PR, and 2.6% in those with low SGM-PR. The proportion of LGBTs is high for those 20–29 and 30–39 years old, for individuals living in households with one member, for individuals who were not working, and for individuals with 10 years or less of residential period. Although the difference in the proportion of LGBTs was small between males and females (sex at birth) and among educational attainment levels, it was slightly higher for men and for individuals with middle level of educational attainment.

Table 2.5 presents the estimated ORs and 95% CIs for LGBTs. The OR of the explanatory variable as a geographical factor was statistically significant both for

Table 2.3 SGM-PRs and
F-PRs by ward

Ward	OCRS		BRR
	SGM-PR	F-PR	F-PR
Kita	2.33	0.74	0.78
Miyakojima	0.91	1.04	0.59
Fukushima	0.71	0.16	0.36
Konohana	1.34	0.00	0.63
Chuo	0.99	0.99	1.70
Nishi	1.67	0.80	0.90
Minato	0.78	0.92	0.78
Taisho	0.27	0.00	0.50
Tennoji	0.32	0.11	1.19
Naniwa	3.82	2.17	2.86
Nishiyodogawa	0.51	0.11	0.79
Yodogawa	1.30	1.72	0.74
Higashiyodogawa	0.32	0.89	0.82
Higashinari	0.70	1.17	1.65
Ikuno	1.49	7.49	5.62
Asahi	0.62	0.99	0.50
Joto	0.80	0.94	0.61
Tsurumi	0.85	0.70	0.31
Abeno	1.21	0.12	0.60
Suminoe	0.28	0.55	0.62
Sumiyoshi	0.44	0.22	0.61
Higashisumiyoshi	1.24	0.95	0.56
Hirano	1.00	0.14	0.80
Nishinari	0.68	3.43	2.12

Note OCRS data are adjusted using sampling weight
Source OCRS, BRR

Model 1 (OR: 18.44, 95% CI: [5.00–71.34]), which included only the explanatory variable, and for Model 2 (OR: 2.38, 95% CI: [1.56–3.57]), which added the control variable. However, the explanatory variable for Model 3, which added interaction terms to Model 2, had an OR of 3.36 (95% CI: [0.40–19.31]) with marginal significance.

Control variables showed relatively similar tendencies between Model 2 and Model 3. For Model 3, while those 20–29 years old (OR: 4.39, 95% CI: [2.26–9.35]) and respondents living in households with one member (OR: 1.68, 95% CI: [1.00–2.74]) tend to more frequently be LGBTs, respondents with white-collar occupations (OR: 0.56, 95% CI: [0.32–0.96]) and gray-collar occupations (OR: 0.55, 95% CI: [0.30–0.98]) were less likely to be LGBTs than respondents who were not working.

Table 2.4 Basic characteristics of respondents and distributions of LGBTs

	Percentage	LGBTs	
		Average	s.d
Geographical factor			
3 wards with high SGM-PR	12.7	0.072	0.259
21 wards with middle or low SGM-PR	87.3	0.026	0.161
Sex at birth			
Male	49.9	0.036	0.185
Female	50.1	0.029	0.168
Age			
20–29	25.3	0.062	0.241
30–39	24.9	0.043	0.203
40–49	28.2	0.011	0.103
50–59	21.5	0.013	0.115
Size of household			
1	19.0	0.061	0.240
2 or more	81.0	0.025	0.158
Education			
Low	32.1	0.029	0.169
Middle	26.0	0.038	0.191
High	41.9	0.031	0.174
Occupation			
White-collar	46.7	0.027	0.163
Grey-collar	25.7	0.030	0.170
Blue-collar	10.1	0.034	0.183
Not working	17.6	0.048	0.214
Years of residence			
Less than 10 years	29.8	0.048	0.213
11 years or more	70.2	0.026	0.158
Total	100.0	0.032	0.177

Notes "s.d." is standard deviation. Data are adjusted using sampling weight
Source OCRS

Regarding migration, individuals with fewer years of residence had an OR of 1.51 (95% CI: [0.96–2.47]) with marginal significance.

Table 2.5 ORs and 95% CIs for LGBTs by logistic regression analysis

	Model 1		Model 2			Model 3		
	ORs	95% CIs	ORs	95% CIs		ORs	95% CIs	
Geographical factor (Ref. 21 wards with middle or low SGM-PR)								
3 wards with high SGM-PR	18.44	***	(5.00–71.34)	2.38	***	(1.56–3.57)	3.36	(0.40–19.31)
Sex at birth (Ref. Male)								
Female			0.73	(0.50–1.06)	0.87	(0.56–1.34)		
Age (Ref. 50–59)								
20–29			4.60	***	(2.47–9.23)	4.39	***	(2.26–9.35)
30–39			3.24	***	(1.71–6.61)	1.73	(0.80–3.96)	
40–49			0.85	(0.38–1.92)	0.92	(0.39–2.19)		
Size of household (Ref. 2 or more)								
1			2.15	***	(1.43–3.19)	1.68	*	(1.00–2.74)
Education (Ref. Low)								
Middle			1.14	(0.70–1.83)	1.17	(0.67–2.05)		
High			0.69	(0.44–1.10)	0.75	(0.44–1.28)		
Occupation (Ref. Not-working)								
White-collar			0.58	*	(0.36–0.93)	0.56	*	(0.32–0.96)
Grey-collar			0.53	*	(0.32–0.88)	0.55	*	(0.30–0.98)
Blue-collar			0.74	(0.37–1.41)	0.85	(0.40–1.71)		
Years of residence (Ref. 11 years or more)								
Less than 10 years			1.16	(0.78–1.72)	1.54	(0.96–2.47)		
Interaction terms								

(continued)

Table 2.5 (continued)

	Model 1		Model 2		Model 3	
	ORs	95% CIs	ORs	95% CIs	ORs	95% CIs
Female × 3 wards with high odds ratios					0.56	(0.24–1.30)
20–29 × 3 wards with high odds ratios					0.87	(0.18–6.35)
30–39 × 3 wards with high odds ratios					4.64	(0.99–33.23)
40–49 × 3 wards with high odds ratios					0.47	(0.04–4.95)
1 × 3 wards with high odds ratios					1.77	(0.75–4.32)
Middle × 3 wards with high odds ratios					0.62	(0.20–1.90)
High × 3 wards with high odds ratios					0.67	(0.23–1.98)
White-collar × 3 wards with high odds ratios					0.88	(0.29–2.79)
Grey-collar × 3 wards with high odds ratios					0.83	(0.25–2.85)
Blue-collar × 3 wards with high odds ratios					0.72	(0.10–4.12)
Less than 10 years × 3 wards with high odds ratios					0.53	(0.21–1.33)
Constant	0.01		0.02		0.02	
n	3992		3992		3992	
− 2 log likelihood	184.2	***	107.86	***	135.96	***
Degree of freedom	1		12		23	

Notes * $p < 0.05$, ** $p < 0.01$, *** $p < 0.001$. OR = Odds ratio; CI = Confidence interval
Source OCRS

2.4   Discussion

1.   Validity of analysis method

The difficulty of painting statistical features of LGBTs using social surveys has already been pointed out (Anderson et al. 2006). This has also been true in clarifying their geographic distribution, mainly due to the limited size of the target groups. Therefore, this study attempted to use weighted samples and then refer to the samples of foreign residents, on which sufficient statistical data and investigations are available.

The results of the tabulation of weighted and unweighted sample statistics demonstrate that the use of the sampling weight makes it possible to correct the distortion of representativeness in the results of OCRS to a certain extent. However, it was also true that the correction was not perfect. In fact, there were some variables for which using the sampling weight did not work sufficiently. The main reason for this is nonresponse bias, which occurs when respondents who refuse to participate in a survey are systematically different from those who participate. The statistics gaps between the weighted samples and the Population Census show that individuals who resided in households with one member, those had a low level of educational attainment, and those not working and with blue-collar occupations were likely to be non-responses.

As for the population distribution by ward, the results of the OCRS and the BRR were nearly consistent as shown by the results of the analysis of foreign residents in Table 2.3. Therefore, the results of the OCRS could be used to examine any small population group in the distribution among wards. However, considering that the F-PR values for each ward were different between the OCRS and the BRR, and that there were some wards where the F-PR values of the OCRS and the BRR were clearly inconsistent, the F-PR values for the 24 wards should only be used for dividing wards into a few categories. Hence, we classified 24 wards into two categories based on whether LGBTs were concentrated in the ward.

Accordingly, despite being a slightly less rigorous approach than that under ideal conditions, the analysis method used in this study to demonstrate the geographic distribution of small population groups based on a social survey could be useful. In addition to geographic distribution, this approach could also be applied to examine various socioeconomic aspects of small population groups quantitatively. This means that the quantitative investigation of small population groups using sampling surveys remains a significant but feasible challenge.

2.   Geographic distribution of LGBTs in Osaka City

As shown in Table 2.3, the SMR-PR values vary from 3.82 for Naniwa Ward to 0.27 for Taisho Ward. This means that LGBTs tend to concentrate in certain wards. However, as shown by Model 3 in Table 2.5, the explanatory variable had an OR with marginal significance, which was estimated using a logistic regression model. Therefore, although there is an apparent tendency for LGBTs to concentrate in certain wards in Osaka City, non-LGBTs with demographic and socioeconomic attributes similar to those of LGBTs also tend to concentrate in these wards. These results suggest that there is no specific mechanism peculiar to the geographic concentration of LGBTs. Nevertheless, it should be noted that the magnitude of the explanatory variable's CIs in the results of logistic regression analysis was large. This may reflect the likely instability of the estimation model due to the small sample size of LGBTs in the OCRS.

For reference, a regression analysis similar to that for LGBTs was performed for foreign residents (details of the results are omitted), and significant associations were seen for the explanatory variable of Model 1 (OR 4.65, 95% CI: [3.29–6.56]), which included only the explanatory variable, of Model 2 (OR 4.02, 95% CI: [2.81–5.75]), which added the control variables to Model 1, and of Model 3 (OR 18.44, 95% CI: [5.00–71.34]), which added interaction terms to Model 2. Although the range of results for 95% CIs of the explanatory variable was considerably wider in Model 3 than in Model 1 and Model 2, unlike in the case of LGBTs, a geographical factor that led to concentration in specific wards appeared to exist in the case of foreign residents.

In this study, when wards were used for the analysis as the geographic units, the results suggest there was no geographical factor that led to geographic concentration of LGBTs. These results, however, do not eliminate the possibility that there are places well suited for sharing identities, culture, and comfortable living environments, which has a certain kind of influence on LGBTs in deciding where to live. In Naniwa Ward and Kita Ward, which have particularly high SMR-PR as shown in Table 2.3, areas with some form of spatial concentration of gay venues are found. In accordance with the developmental model for gay neighborhoods that thrived in Western countries (e.g., Collins and Drinkwater 2017), the concentration of service enterprises such as bars, clubs, and restaurants for LGBTs serves as a trigger for the formation of gay villages. Therefore, a similar phenomenon may be emerging in Osaka City, where LGBTs tend to also concentrate in certain areas in specific wards. However, it has been pointed out that this developmental model may not apply in the non-Western world (Yue and Leung 2017; Suzaki 2019b). Therefore, further studies are required to determine whether LGBTs are concentrated in wards and areas within particular wards in Osaka City and, if so, how these concentrated wards and areas are built. Such studies should be carried out by a range of contrasting methods, such as the analysis of a large sampling survey, a field survey in these areas, and interviews focusing on LGBTs' decisions on their place of residence.

As for migration, which is assumed to be associated with the geographic distribution of population, individuals with fewer years of residence had an OR of 1.51 (95% CI: [0.96–2.47]) with marginal significance using Model 3, as shown in Table 2.5, although significant associations were seen for the corresponding results of the analysis on foreign residents conducted as reference (OR 4.62, 95% CI: [2.69–8.10]). This suggests that there is no difference between LGBTs and non-LGBTs, in contrast to between Japanese and foreigners, in their tendency to migrate into Osaka City. Studies on the migration of LGBTs have revealed the existence of various migration processes and factors associated with their life courses, as well as migration streams toward large cities (Cooke and Rapino 2007; Lewis 2014). The OCRS did not include any question on migration, so further studies are necessary to examine whether migration affects the geographic distribution of LGBTs.

2.5 Conclusion

This study used microdata from one of the first representative random sampling surveys, called "Survey on Diversity of Work and Life, and Coexistence among the Residents of Osaka City" (OCRS). This was conducted in January 2019 with the cooperation of the Osaka City government in Japan by the research team of the project "Demography of Sexual Orientation and Gender Identity: Building a Foundation for Research in Japan," which includes the author (Principal Investigator: Saori Kamano); the project was established to examine the geographic distribution of LGBTs, including gay, lesbian, bisexual, and transgender people in Osaka City. In consideration of the small sample numbers of LGBTs in the OCRS, this study attempted to use weighted samples and to refer to the samples of foreign residents, on which sufficient statistical data and investigations are available.

The results demonstrated that the analysis method adopted in this study could be effective and that using a sampling survey to examine the geographic distribution of small population groups could be practicable. Regarding the geographic distribution of LGBTs, although there was an apparent tendency for them to concentrate in certain wards of Osaka City, it was found through logistic regression analysis that non-LGBTs with demographic and socioeconomic attributes similar to those of LGBTs also tended to concentrate in these wards. In other words, it could be assumed that there is no specific mechanism peculiar to the geographic concentration of LGBTs. Therefore, we conclude that the uneven distribution of LGBTs in Osaka City may not reflect the geographical context of LGBTs' concentration but instead mirror general variations in the population composition. Although further studies are necessary, by providing initial findings regarding the geographic distribution of LGBTs in Japan using social surveys quantitatively, this study may assist researchers and others to better understand the socioeconomic situations of LGBTs in Japan.

References[3]

Anacker KB, Morrow-Jones HA (2005) Neighborhood factors associated with same-sex households in US cities. Urban Geography 26(5): 385–409. https://doi.org/10.2747/0272-3638.26.5.385

Anderson G, Noack T, Seierstad A et al (2006) The demographics of same-sex marriages in Norway and Sweden. Demography 43(1): 79–98. https://doi.org/10.1353/dem.2006.0001

Baumle AK (ed) (2013) International handbook on the demography of sexuality. Springer, Netherland

---

[3] (J): Written in Japanese.

Black D, Gates G, Sanders S et al (2000) Demographics of the gay and lesbian population in the United States: Evidence from available systematic data sources. Demography 37(2): 139–154. https://doi.org/10.2307/2648117

Blank Y, Rosen-Zvi I (2012) The geography of sexuality. North Carolina Law Review 90(4): 955–1026. https://scholarship.law.unc.edu/nclr/vol90/iss4/2

Collins A, Drinkwater S (2017) Fifty shades of gay: Social and technological change, urban deconcentration and niche enterprise. Urban Studies 54(3): 765–785. https://doi.org/10.1177/004209 8015623722

Cooke TJ, Rapino M (2007) The migration of partnered gays and lesbians between 1995 and 2000. The Professional Geographer 59(3): 285–297. https://doi.org/10.1111/j.1467-9272.2007. 00613.x

Duncan S, Smith D (2006) Individualisation versus the geography of 'new' families. Journal of the Academy of Social Sciences 1(2): 167–189. https://doi.org/10.1080/17450140600906955

Firth D (1993) Bias reduction of maximum likelihood estimates. Biometrika 80(1): 27–38. https://doi.org/10.1093/biomet/80.1.27

Fukumoto T (2010) Tokyo oyobi Osaka ni okeru zainichi gaikokujin no kukanteki segurigeishon no henka: "Orudokama" to "nyukama" kan no sai ni chakumoku shite (Changes in spatial segregation of foreigners in Tokyo and Osaka: Differences between "Oldtimers" and "Newcomers"). Chirigaku Hyoron Series A (Geographical review of Japan Series A) 83(3): 288–313. https://doi.org/10.4157/grj.83.288 (J)

Fukumoto T. (2018a) Zainichi chosenjin jigyosho no kukanteki bunpu to syujuchiku tono kanrensei: 1980 nendai iko no Osaka wo jirei ni (The relationship between the spatial distribution of ethnic entrepreneurs and ethnic residential clusters: A case of Koreans in Osaka after the 1980s). Keizai Chirigaku Nenpo (Annals of the Association of Economic Geographers) 64(3): 194–216. https://doi.org/10.20592/jaeg.64.3_194 (J)

Fukumoto T (2018b) Nihon no toshi ni okeru esunikku segurigeishon kenkyu no doko (Ethnic segregation studies in Japan: Retrospect and prospect). Toshi Chirigaku (Urban Geography) 13: 77–91. https://doi.org/10.32245/urbangeography.13.0_77 (J)

Gates GJ (2013) Geography of the LGBT population. In: Baumle AK (ed) International handbook on the demography of sexuality. Springer, Netherland, p 229–242

Goldie X (2018) Together, but separate: Neighborhood-scale patterns and correlates of spatial segregation between male and female same-sex couples in Melbourne and Sydney. Urban Geography 39(9): 1391–1417. https://doi.org/10.1080/02723638.2018.1456030

Heinze G, Schemper M (2002) A solution to the problem of separation in logistic regression. Statistics in Medicine 21(16): 2409–2419. https://doi.org/10.1002/sim.1047

Hiramori D, Kamano S (2020) Asking about sexual orientation and gender identity in social surveys in Japan: Findings from Osaka City Residents' survey and related preparatory studies. Jinko Mondai Kenkyu (Journal of Population Problems) 76: 443–466. https://doi.org/10.31235/osf. io/w9mjn

Kamano S. (2018) Seiteki mainoritei wo meguru ryoteki data: Daibasitei suishin no bunmyaku ni okeru ryogisei (Quantitative data on sexual minorities: Issues and potentials in the promotion of diversity). Joseigaku (Women's Studies) 26: 22–37. https://doi.org/10.50962/wsj.26.0_22 (J)

Kamano S, Ishida H, Iwamoto T et al (2019) Osaka shimin no hatarakikata to kurashi no tayosei to kyosei ni kansuru anketo hokokusyo (tanjun shukei kekka) (Survey on diversity of work and life, and coexistence among the residents of Osaka City: Report based on percent frequency tables). http://www.ipss.go.jp/projects/j/SOGI. Accessed 30 Sep 2022 (J)

Kamiya H (2018) Beshikku toshisyakai chirigaku (Urban social geographies). Nakanishiya, Kyoto (J)

Kanai JM, Kenttamaa-Squires K (2015) Remaking south beach: Metropolitan gayborhood trajectories under homonormative entrepreneurialism. Urban Geography 36(3): 385–402. https://doi.org/10.1080/02723638.2014.970413 (J)

Lewis NM (2014) Moving "out," moving on: Gay men's migrations through the life course. Annals of the Association of American Geographers 104(2): 225–233. https://doi.org/10.1080/00045608. 2013.873325

Mitsuhashi J (2018) Shinjuku: "Sei naru machi" no rekisichiri (Historical geographies of hetero and homo sexualities in Shinjuku). Tokyo. Asahi Shimbun Publications, Tokyo (J)

Schroeder CG (2014) (Un)holy Toledo: Intersectionality, interdependence, and neighborhood (trans)formation in Toledo, Ohio. Annals of the Association of American Geographers 104(1): 166–181. https://doi.org/10.1080/00045608.2013.847756

Smart MJ, Whittemore AH (2017) There goes the gaybourhood? Dispersion and clustering in a gay and lesbian real estate market in Dallas TX, 1986–2012. Urban Studies 54(3): 600–615. https://doi.org/10.1177/0042098016650154

Sunagawa H (2015) Shinjuku ni-chome no bunka-jinruigaku: Gay community kara toshi wo manazasu (Ethnographies of Shinjuku ni-chome: Gazing the city from gay communities). Tarojiro-Sha Editus, Tokyo (J)

Suzaki S (2019a) "Shinjuku ni-chome" chiku ni okeru gei dansei no basyo imeji to sono henka (Image of Shinjuku ni-chome as a gay neighborhood and its evolution). Chirigaku Hyoron Series A (Geographical Review of Japan Series A) 92(2): 72–87. https://doi.org/10.4157/grj. 92.72 (J)

Suzaki S (2019b) Shinjuku ni-chome ni okeru gei dyisutorikuto no kukanteki tokucho to sonzoku joken (Spatial characteristics and existence conditions of the gay district in Shinjuku 2-chome, Tokyo). Toshi Chirigaku (Urban Geography) 14: 16–27. https://doi.org/10.32245/urbangeog raphy.14.0_16 (J)

Wimark T, Östh J (2014) The city as a single gay male magnet? Gay and lesbian geographical concentration in Sweden. Population, Space and Place 20(8): 739–752. https://doi.org/10.1002/ psp.1825

Yue A, Leung HH (2017) Notes towards the queer Asian city: Singapore and Hong Kong. Urban Studies 54(3):747–764. https://doi.org/10.1177/0042098015602996

Chapter 3
Future Prospects of Regional Population in Japan

Shiro Koike

Abstract This chapter outlines the methods and results of the "Regional Population Projections for Japan: 2015–2045" ("RPJ2018" hereinafter) drafted by the National Institute of Population and Social Security Research, and it presents a view of foreign population and centralization in the Tokyo metropolitan area. These factors could be influenced through the introduction of national or local policies, among other measures. RPJ2018 presents high population decline rates, especially in non-metropolitan areas, which indicates that a greater part of this decline will be the result of a natural decrease as the pressures of nationwide population decline intensify. As such, it appears very difficult to halt the population decline among the foreign population located in non-metropolitan areas, given that this group is more concentrated in metropolitan areas (compared with the general Japanese population) coupled with a drop in the fertility rate throughout the foreign population. On the other hand, the over-concentration in the Tokyo metropolitan area will continue in the long run because the out-migration mobility (from the Tokyo metropolitan area) will decline due to the percentage increase in the number of people born in the Tokyo metropolitan area, along with people whose parents were also born in the same district. Since any region's ability to achieve a population increase via in-migration will be limited to very small areas, it can be argued that all regional plans need to be formulated on the premise that population decline—mainly as a result of a natural decline—is a given.

Keywords Regional population projections for Japan · Natural decrease · Foreign residents · Over-concentration in Tokyo metropolitan area

S. Koike (✉)
Department of Population Structure Research, National Institute of Population and Social Security Research, Chiyoda-ku, Tokyo, Japan
e-mail: koike-shiro@ipss.go.jp

© The Author(s), under exclusive license to Springer Nature Singapore Pte Ltd. 2023
Y. Ishikawa (ed.), *Japanese Population Geographies II*,
Population Studies of Japan, https://doi.org/10.1007/978-981-99-2076-1_3

37

3.1 Introduction

According to the National Institute of Population and Social Security Research (hereinafter, "IPSS") publication titled Regional Population Projections for Japan: 2015–2045 (hereinafter, "RPJ2018"), 1,588 municipalities (94.4% of total municipalities) will experience a decrease in total population in the 30-year period from 2015 to 2045. Of these, 334 (approximately 20% of total) are projected to see their population decrease to less than half. Furthermore, aging of the population is also advancing; it is projected that by 2045, the proportion of the population aged 65 and over (hereinafter, population aged 65 +) will exceed 50% in 465 municipalities, or almost 30% of all municipalities. In response to such rapid decline and super-aging of the population, the Japanese government has focused on regional revitalization as a major policy, with the aim of creating a decentralized and sustainable society that takes advantage of each region's unique characteristics. Moreover, the government is trying to rectify over-concentration in the Tokyo metropolitan area and to stop the decline of regional populations. In 2014, the government formulated the Town/People/Work Creation Long-Term Vision (hereinafter, "Long-Term Vision"), which visualizes the current state and the ideal future state of Japan's population, and the Town/People/Work Creation Comprehensive Strategies (hereinafter, "Comprehensive Strategies"), which summarize the basic directions and concrete initiatives of policy goals and measures. In this connection, local governments have been mandated to formulate a Municipal Population Vision and a Municipal Version of the Comprehensive Strategies. The second phase of the Long-Term Vision and the Comprehensive Strategies was formulated in 2019. However, it is not easy to halt the decline in regional population, which is caused by the long-standing low fertility rate and the outflow of the young population. Today, many regions are already facing a number of problems associated with declining population, such as significant decreases in tax revenues, difficulties in maintaining infrastructure, and the deterioration of local communities. With the certainty of future decline and aging of the population throughout almost all of Japan, what kind of strategies should each region adopt toward implementing its policies?

The 24th IPSS Annual Seminar, where the author gave the keynote speech, was organized to address the above issues. This paper is a summary of that keynote speech. It discusses the future prospects of the regional population, which are indispensable in considering the future of local communities, as well as the trends in the foreign population and the over-concentration of population in the Tokyo metropolitan area, which are major issues related to regional population.

3.2 Future Prospects of Regional Population According to RPJ2018

IPSS conducts projections on the nationwide and regional populations based on each Population Census. The latest regional population projections are included in RPJ2018, which is based on the 2015 Population Census. The projections cover a period of 30 years, from 2015 to 2045, and a total of 1,799 regions as of March 1, 2018, including one prefecture (Fukushima Prefecture)[1] and 1,798 municipalities (i.e., 23 wards of Tokyo (special wards), 128 wards of 12 ordinance-designated cities,[2] 766 cities, 713 towns, and 168 villages). Projections were carried out using the cohort component method, which is currently the most widely adopted method worldwide.

RPJ2018 was also carried out from the viewpoint of "projection," as is the case for all future population projections conducted by IPSS. In other words, the projection results are based on the assumption that the recently observed trends in birth, death, and migration will generally continue in the future. Therefore, it should be noted that socioeconomic changes possibly occurring in the future (e.g., regional economic conditions, transportation infrastructure development, location of facilities, development of residential areas) and the changes in population movements caused by regional policies are not factored in the projection results.

According to the results of projections using the median birth and death assumptions from the Population Projections for Japan: 2016–2065 (hereinafter, "PPJ 2017"), the total population index for 2045 is 83.7, based on the total population of Japan in 2015 as 100. By prefecture, only Tokyo (100.7) barely exceeds the baseline, while the remaining 46 prefectures are below 100, with Akita Prefecture (58.8), showing the highest rate of population decline, projected to decrease by more than 40% over the 30-year period (Fig. 3.1, left). By metropolitan and non-metropolitan areas,[3] the indices are 93.8 for the Tokyo metropolitan area, 87.3 for the Nagoya metropolitan area, 81.2 for the Osaka metropolitan area, and 77.7 for the non-metropolitan areas. While it is projected that the population of the Tokyo metropolitan area will remain nearly stagnant over the 30-year period, the population of the Osaka metropolitan area will decrease at a faster rate than that of the entire country.

The aging of the population is also progressing steadily. According to PPJ2017, the proportion of the population aged 65 + will rise from 26.6% in 2015 to 36.8% in 2045. The proportion of the population aged 65 + is projected to exceed 30% in

---

[1] For Fukushima Prefecture, due to the impact of the nuclear accident at the Fukushima Daiichi Nuclear Power Plant following the Great East Japan Earthquake on March 11, 2011, it has been extremely difficult to forecast municipal population trends and future changes; consequently, the population projection was made for the prefecture as a whole.

[2] Sapporo, Sendai, Chiba, Yokohama, Kawasaki, Nagoya, Kyoto, Osaka, Kobe, Hiroshima, Kitakyushu, and Fukuoka.

[3] Tokyo metropolitan area includes Saitama, Chiba, Tokyo, and Kanagawa; Nagoya metropolitan area includes Gifu, Aichi, and Mie; and Osaka metropolitan area includes Kyoto, Osaka, Hyogo, and Nara.

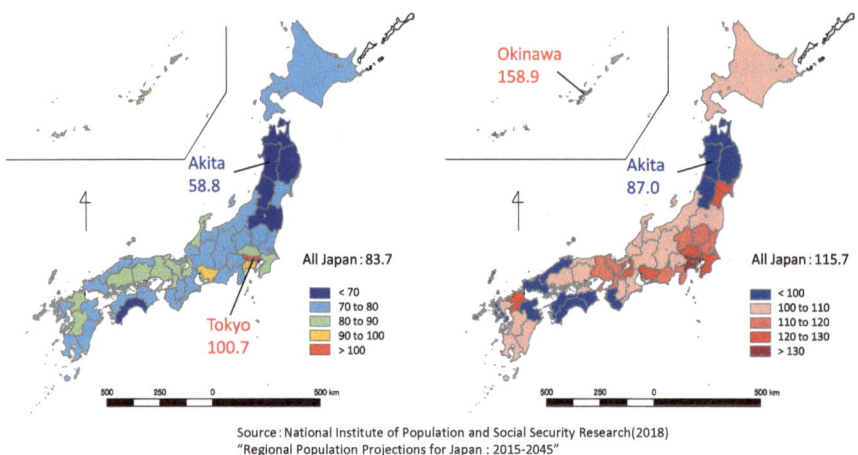

Source: National Institute of Population and Social Security Research (2018)
"Regional Population Projections for Japan : 2015-2045"

Fig. 3.1 Index of total population in 2045 (2015 = 100) (left) and 65 + population in 2045 (2015 = 100) (right)

all prefectures by 2045. In Akita Prefecture, which has the highest proportion of the population aged 65 +, this rate is projected to reach 50.1%, i.e., one elderly person for every two persons. In terms of the rate of the elderly to the total population, the non-metropolitan areas have conspicuously higher rates; however, in terms of the absolute number of the elderly population, the increase in the metropolitan areas is larger. Figure 3.1 (right) shows the index of the population aged 65 + by prefecture in 2045 based on the population aged 65 + in 2015 as 100. With the exception of Okinawa Prefecture, which has the highest rate of increase, all prefectures belonging to major metropolitan areas or prefectures with regional central cities have high rates of increase across the board. On the other hand, the 12 prefectures belonging to non-metropolitan areas will see a decrease in the population aged 65 + over the 30-year period. These regional changes in the population aged 65 + become clearer when we look at the distributions of the population turning 65 and over by 2045 and those people who were aged 35 to 64 in 2015. Plotting the ratio of the population aged 35 to 64 to the population aged 65 + in 2015 on the horizontal axis and the population aged 65 + in 2045 based on the population aged 65 + in 2015 as 100 on the vertical axis (Fig. 3.2), we see that the prefectures line up in a nearly straight line. The distribution of the population aged 35 to 64 largely dictates the changes in the population aged 65 + over the 30-year period because there is less movement across prefectures among people aged 35 and older, and they tend to live in the same prefecture until they reach 65 and above.

Looking at the results by municipality, we see that the regional differences are naturally larger than those among prefectures, with the highest total population index in Chuo Ward, Tokyo (134.9) and the lowest in Kawakami Village, Nara Prefecture (20.6) in 2045. From the total population index by municipality based on population size in 2015, it is clear that municipalities with smaller populations tend to have a

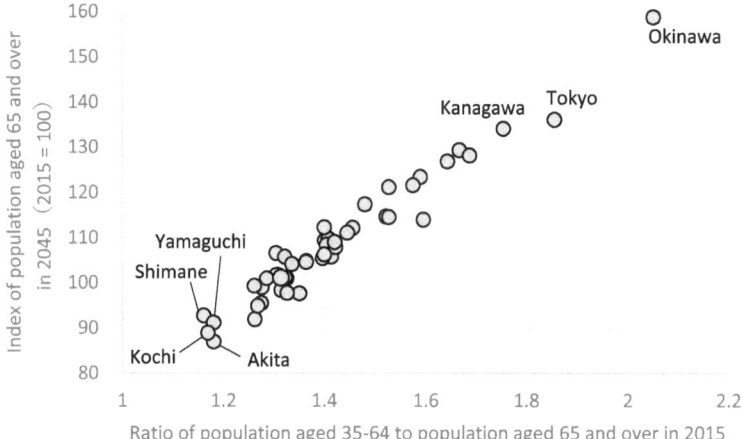

Source : National Institute of Population and Social Security Research(2018)
"Regional Population Projections for Japan : 2015-2045"

Fig. 3.2 Relationship of ratio of population aged 35–64 to population aged 65 + in 2015 and the index of population aged 65 + in 2045 by prefecture

higher rate of population decline. The total population of the 23 wards of Tokyo and ordinance-designated cities will decrease by less than 5% in 2045, while the total population of municipalities with a population of less than 10,000 is projected to decrease in 2045 to about half of the population in 2015. In addition, while only 15 (0.9%) of the municipalities had more than 50% of their population aged 65 + in 2015, this number is expected to reach 465 (27.6%) in 2045, pointing to a considerable increase in aging in underpopulated areas.

On the other hand, in terms of the absolute number of the elderly population, there is a remarkable increase in the elderly population in metropolitan areas, as in the case of prefectures. Looking at the changes in the population aged 75 and over from 2015 to 2045 around the Tokyo metropolitan area, we see that municipalities that will experience a significant increase in this population range are spread widely in the suburban areas surrounding central Tokyo, with some municipalities seeing this population more than double in the 30-year period. This clearly reflects the migration pattern among the generation older than the 1960 birth cohort (the so-called first suburban generation), who have moved to the central part of the Tokyo metropolitan area for higher education or employment and then moved to the suburbs upon getting married or raising children (Kawaguchi 2007). Reflecting this migration pattern, as of 2015, the population aged 45 to 74 was widely distributed in suburban areas. Thus, assuming few changes in the place of residence, there will be a particularly higher rate of increase in the late elderly population in the suburban districts of metropolitan areas.

Looking at the rate of change in the population between 2015 and 2045 in 1682 municipalities, including those with ordinance-designated cities as a single city as

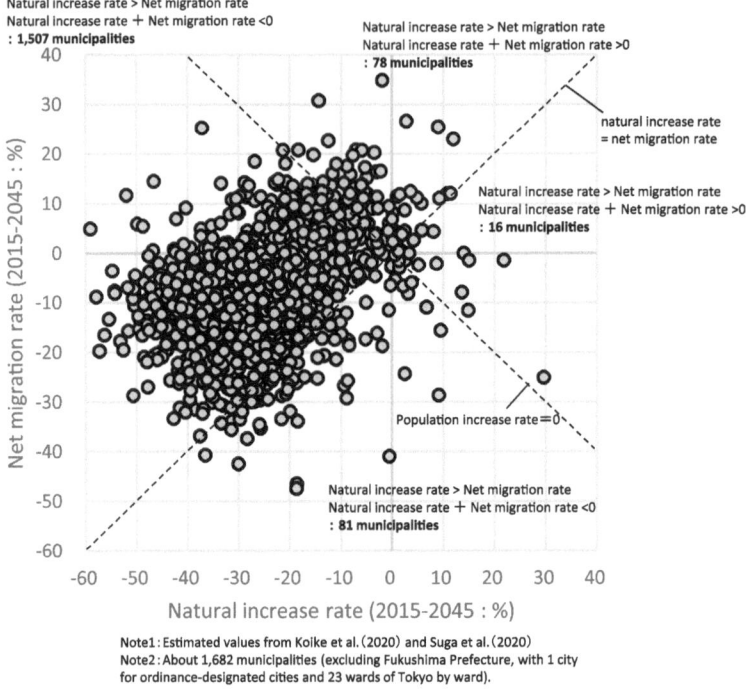

Fig. 3.3 Distribution of natural increase rate and net migration rate by municipalities (2015–2045)

well as the 23 wards of Tokyo as distinct municipalities, in terms of the natural increase rate and the net migration rate (Fig. 3.3),[4] we see that among the 1,588 municipalities with populations decreasing over the 30-year period, up to 1,507 of them have a natural increase rate lower than the net migration rate. This indicates that the majority of the municipalities will experience a decrease in population due to natural decline. For example, in Kawakami Village, Nara Prefecture, which has the lowest total population index in 2045, the natural increase rate over the 30-year period is − 50.7%, which is considerably lower than the net migration rate over the same period (− 28.7%). The proportion of the population aged 65 + in Kawakami Village, according to the 2015 Population Census, is 58.7%, indicating marked aging of the population, which makes the rapid natural decline inevitable due to the population structure there. In non-metropolitan areas, various measures have been taken to halt the outflow of young people, but even if these measures are effective, the decline in the population will not stop because the number of deaths significantly exceeds the number of births.

---

[4] Natural increase rate was calculated by subtracting the number of deaths estimated by Suga et al. (2020) from the number of births estimated by Koike et al. (2020). The net migration rate was calculated by subtracting the natural increase rate from the population increase rate.

3.3 Issues Surrounding Projected Regional Populations

Since, as mentioned above, RPJ2018 was carried out from the viewpoint of "projection," changes in demographic trends, including migration, that occur during the projection period are basically not taken into account. Therefore, the actual population may naturally deviate from the projections due to the formulation of new government policies and other factors. In the following, we focus on the foreign population and the over-concentration in the Tokyo metropolitan area, factors that are likely to change for considering future policies, and add a brief discussion after looking at the data related to recent trends and future prospects.

1. Foreign population

The total population of Japan peaked around 2008 and has declined since then. The country's foreign population, however, has been increasing almost constantly. It has increased significantly from approximately 783,000 in 1980 to approximately 2,731,000 in 2018 (Fig. 3.4, left). By nationality, the highest foreign population in 2018 was from China, followed by Korea, Vietnam, the Philippines, and Brazil, which, taken together, account for 77% of the total foreign population. The new "Specified Skilled Worker" visa, established in 2019 to expand the acceptance of foreign human resources, has led to high expectations that the increase in the population of foreigners may somehow curb the decline in Japan's overall population.

The differences in regional distribution and age structure are among the characteristics that distinguish the foreign population from the Japanese population. First, regarding regional distribution, in 2019 the highest ratio of foreign population to total population was in Tokyo (4.0%) and the lowest was in Akita Prefecture

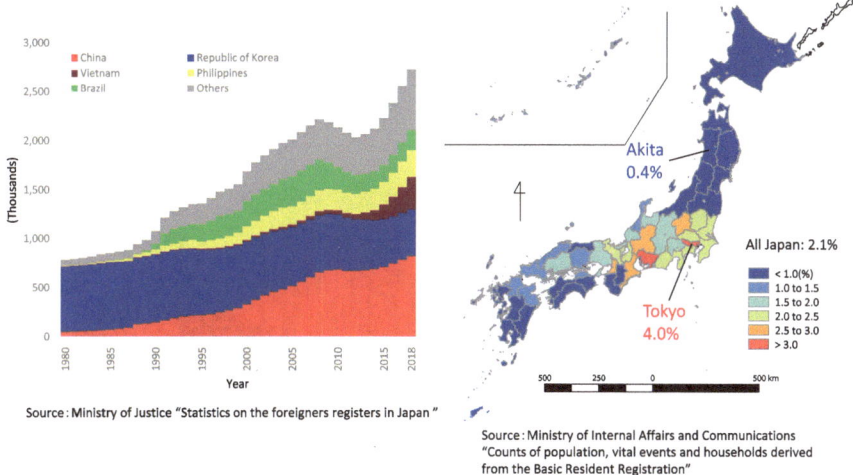

Source: Ministry of Justice "Statistics on the foreigners registers in Japan"

Source: Ministry of Internal Affairs and Communications "Counts of population, vital events and households derived from the Basic Resident Registration"

Fig. 3.4 Foreign residents in Japan by nationality (1980–2018) (left) and percentage of foreign population by prefecture (2019) (right)

(0.4%), according to the "Survey of population, demographics, and number of house-holds based on the Basic Resident Registration" conducted by the Local Administration Bureau, Ministry of Internal Affairs and Communications (Fig. 3.4, right). The foreign population was markedly more concentrated in metropolitan areas than was the Japanese population. As of 2019, the Tokyo metropolitan area accounted for 41.0% of the total foreign population (28.5% for Japanese), while the three major metropolitan areas together accounted for 70.1% (51.7% for Japanese). With the exception of some regions, the proportion of foreign residents in regional areas remained low.

Regarding age structure, Fig. 3.5 shows population pyramids of the foreign population and the Japanese population in 2019. In Japan, the generations after the second baby boom have shrunk almost constantly, forming a so-called pot-shaped pyramid, while the foreign population forms a "tree-shaped pyramid." By age group, the proportion of foreigners aged 15 to 64 is very high, while the proportions of the population aged 65 + and aged 0 to 14 are very low. Moreover, the population pyramid for foreigners has maintained almost the same shape over time and has not changed much in recent years. This reflects the fact that foreign nationals, such as international students and technical intern trainees, mainly stay in the country on a short-term basis. In addition, at least for now, their contribution to the birth rate of the next generation cannot be considered significant. As shown by Nakagawa et al. (2018), the total fertility rate for foreign nationals peaked around 2000 and has since declined rapidly. In fact, it has fallen below the total fertility rate for Japanese nationals since 2010. In particular, in the 1990s, many women came to Japan as "foreign brides," mainly in the regional areas, as one of the measures to address the lack of successors. In recent years, however, the proportion of those who enter Japan alone and then leave the country alone has been increasing.

Given these circumstances, it is clear that it would be very difficult for the foreign population to stop the population decline in the long term, although they may temporarily alleviate it in many regional areas. Naturally, the foreign population will also be largely affected by future policies. Presently, however, its contribution to

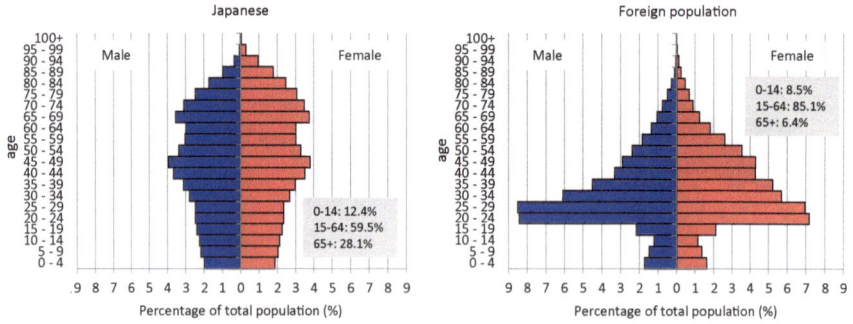

Fig. 3.5 Population pyramids of Japanese and foreign populations living in Japan (2019)

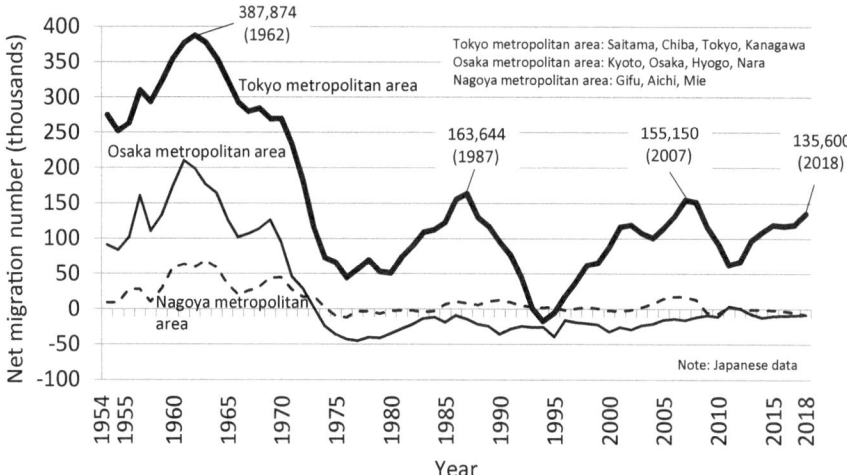

Fig. 3.6 Net migration numbers in three major metropolitan areas (1954–2018)

natural increase is small due to the low fertility rate. Therefore, it is extremely difficult to rely on the foreign population to compensate for the natural decline of the Japanese population, which is expected to continue to worsen even in the metropolitan areas.

2. Over-concentration in Tokyo metropolitan area

One of the major features of the recent domestic migration is the over-concentration of population in the Tokyo metropolitan area. As shown in Fig. 3.6, from the 1980s onward, among the three major metropolitan areas, only the Tokyo area has exhibited a distinctive positive net migration trend.

Aiming to correct this trend of over-concentration in the Tokyo metropolitan area, the government adopted "regional revitalization" as a major policy in 2014. Despite the various measures that have been implemented mainly in the regional areas, however, positive net migration in the Tokyo area in 2018 reached approximately 136,000 (Japanese residents), close to the peak during the "bubble" period. Separating the net migration into in-migration and out-migration, we can see that the decreasing trend in out-migration is particularly conspicuous. Will the over-concentration in the Tokyo metropolitan area continue in the future? In this paper, we used the results of the 8th National Survey on Migration conducted by IPSS in 2016 to examine changes in the distribution of birthplaces (Koike and Shimizu 2020).

Figure 3.7 shows a very simplified model of the changes in place of residence and birthplace for a hypothetical family. In this family, it is assumed that the second generation (assumed to be the first baby boomer generation) moved from somewhere outside of the Tokyo metropolitan area to the Tokyo metropolitan area, while the other generations did not move between the Tokyo metropolitan area and an outer area.

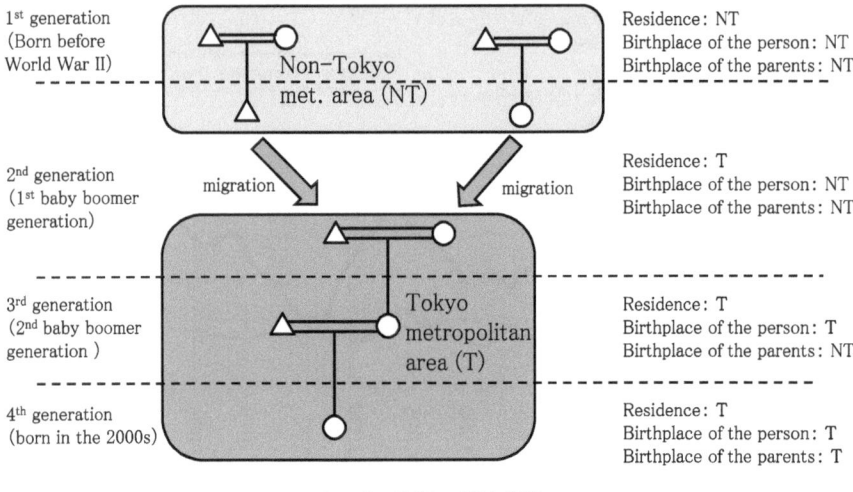

1st generation (Born before World War II) — Non-Tokyo met. area (NT) — Residence: NT / Birthplace of the person: NT / Birthplace of the parents: NT

2nd generation (1st baby boomer generation) — migration / migration — Residence: T / Birthplace of the person: NT / Birthplace of the parents: NT

3rd generation (2nd baby boomer generation) — Tokyo metropolitan area (T) — Residence: T / Birthplace of the person: T / Birthplace of the parents: NT

4th generation (born in the 2000s) — Residence: T / Birthplace of the person: T / Birthplace of the parents: T

Source: Figure3 in Koike and Shimizu (2020)

Fig. 3.7 Patterns of change in residence and birthplace for a hypothetical family

For the first generation (assumed to be born before World War II), their place of residence and birthplace and the birthplace of their parents are all in an area outside of the Tokyo metropolitan area. For the second generation, who migrated, their birthplace and the birthplace of their parents are both in the area outside of the Tokyo metropolitan area, while their place of residence is in the Tokyo metropolitan area. For the third generation (assumed to be the second baby boomer generation), the birthplace of their parents is in the non-Tokyo met. area, while their birthplace and their place of residence are in the Tokyo metropolitan area. For the fourth generation (assumed to be born mainly in the 2000s), their place of residence, birthplace, and birthplace of their parents are all in the Tokyo metropolitan area. In other words, when migration occurs from a non-Tokyo met. area to the Tokyo metropolitan area, a change in the distribution of place of residence occurs first, followed by a change in the distribution of birthplace in the next generation and, consequently, a change in the distribution of birthplace of the parents in the generation after that. It is a well-known fact that during the period of rapid economic growth, there was a massive migration of the population, mainly among the first-generation baby boomers, from areas outside the Tokyo metropolitan area into the Tokyo metropolitan area, which led to a drastic change in the population distribution. However, the changes in the distribution of birthplace have remained almost unknown because the relevant information could not be obtained from the Population Census and other major statistics. Looking at the birthplace by age group of Tokyo metropolitan area residents in accordance with whether they were born in the Tokyo area, according to the 8th National Survey on Migration, we can see that the rate of people born in the Tokyo metropolitan area is higher among those aged 40 to 44 and 45 to 49, who generally represent the second-generation baby boomers, indicating a change in the distribution of birthplace.[5]

Next, looking at the birthplace of parents by age, specifically for residents born in the Tokyo metropolitan area (Fig. 3.8, top), we see that the rate of those with "both parents born in the Tokyo metropolitan area" is highest among those aged 15 to 19, who generally represent the generation of children of the second-generation baby boomers. Following the change in the distribution of birthplace, the distribution of the birthplace of parents also changed among the younger generation, a phenomenon that is similar to that shown in the simplified model in Fig. 3.7. Likewise, if we look at the current place of residence in accordance with birthplace of parents by age, specifically for residents born in the Tokyo metropolitan area (Fig. 3.8, bottom), we can see that about 1 in 4 people aged 20 or older live in non-Tokyo met. areas for those with both parents born in non-Tokyo met. areas. Even those born in the Tokyo metropolitan area may move out of the Tokyo area if both of their parents were born in non-Tokyo met. areas, resulting in a considerable number of them living outside the Tokyo area. Accordingly, having both parents born in non-Tokyo met. areas increases the likelihood for people born in the Tokyo metropolitan area to make return migrations, resulting in a considerable number of them living outside of the Tokyo area.

On the other hand, the proportion of those living in non-Tokyo met. areas among those with both parents born in the Tokyo metropolitan area is extremely low for all age groups, at about 1–2%. The proportion of non-Tokyo residents among those with either of their parents born in the Tokyo metropolitan area falls between those of the above two groups but is slightly closer to the proportion among those with both parents born in the Tokyo area, at about 5–10%. In other words, while the birthplace of people greatly affects their choice of residence, the birthplace of their parents also plays a major role in that choice. Thus, the recent decrease in the volume of out-migration from the Tokyo metropolitan area mentioned above could be largely attributed to the change in the distribution of birthplace of Tokyo metropolitan area residents.

The concentration of the population in the Tokyo metropolitan area will likely result in the continued increase in the proportion of people born in the Tokyo area and the proportion of those whose parents are both born in the Tokyo area. Likewise, the return migrations out of Tokyo, which thus far have likely accounted for a large part of the migration from the Tokyo metropolitan area to areas outside of it, will decrease. Moreover, the number of people staying in the Tokyo metropolitan area even after graduating from university or finding employment will also increase further. These observations show that changes in migration trends are closely related to demographic attributes, making it highly probable that the over-concentration of population in the Tokyo metropolitan area will continue in the long term due to the decline in out-migration mobility.[6]

---

[5] See Table 2 in Koike and Shimizu (2020).

[6] After the 24th IPSS Annual Seminar, the trends in both international and domestic migration have changed significantly due to the COVID-19 pandemic. International migration, both inward

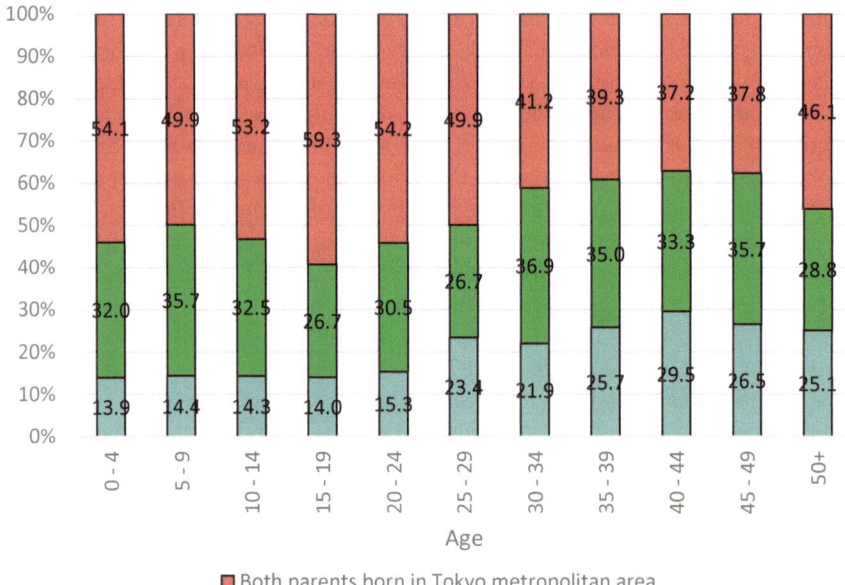

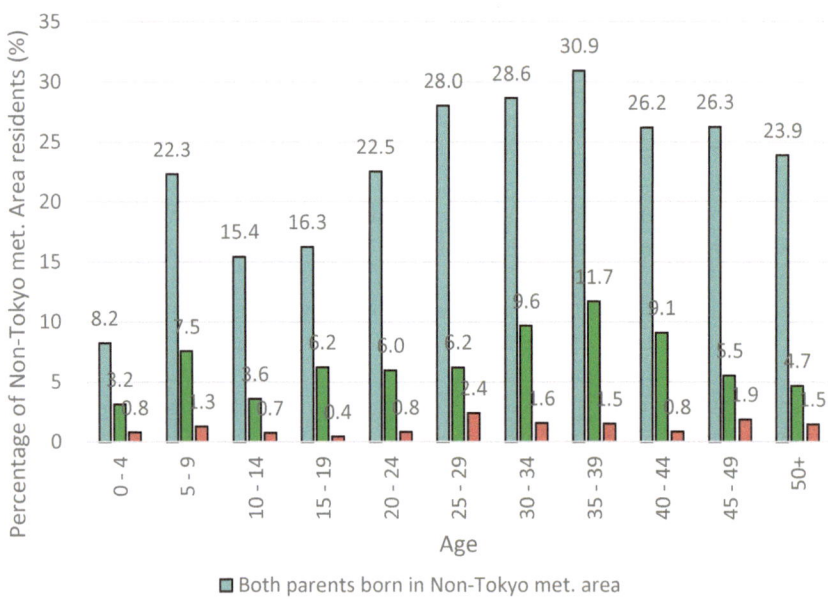

Source: Figures 6 and 7 in Koike and Shimizu (2020)

Fig. 3.8 Distribution of parents' birthplace by age (top) and percentage of non-Tokyo met. area residents by the parents' birthplaces and ages (bottom)

3.4 Conclusion

Presently, the majority of Japan's regions are already experiencing a decline in population, but the main cause of the decline is shifting from negative net migration to natural decline. In particular, in many regional areas, the negative net migration has been mitigated primarily by the fact that there are very few young people who will leave. Meanwhile, the number of births has dropped sharply due to the declining fertility rate and the conspicuous decline in the maternal population. Although attempts have been made to stop the decline in the population of non-metropolitan areas through measures aimed at increasing the foreign population and correcting the over-concentration in the Tokyo metropolitan area, this paper shows that these two goals are difficult to achieve from a demographic point of view.

Since almost all regions are experiencing a natural decline, population growth due to migration will take place in only a few regions. In regard to domestic migration, which accounts for a large part of population movements, an increase in in-migration in one region means an increase in out-migration in another region, making it impossible to maintain the country's population size through net migration in all regions. Even in the Tokyo metropolitan area, where over-concentration is progressing, the volume of in-migration will decrease over the long term due to the decline in the young population in areas outside of the Tokyo metropolitan area, which has thus far supported the population increase. Thus, with its low fertility rate, the population will certainly begin to decline in the near future. Therefore, all regional plans need to be formulated on the premise that population decline—mainly as a result of natural decline—is a given.

References[7]

Kawaguchi K (2007) *Shakai keizai teki jinko zokusei kara mita daitoshi-ken kukankozo no hensen* (Changing metropolitan spatial structures from the viewpoint of socio-economic demography). *Meiji Daigaku Jinbunkagaku Kenkyusho Kiyo* (Memoirs of the Institute of Humanities, Meiji University), 60: 53–76 (J)

---

and outward, has nearly stopped. Likewise, in regard to domestic migration, it has been reported that the trend of over-concentration in the Tokyo metropolitan area has slowed down. As pointed out above, it is currently difficult to predict whether the trend of over-concentration in the Tokyo area will change over the long term because it is accompanied by changes in the distribution of birthplaces of residents. On the other hand, the diversification of workplaces has been progressing due to the relocation of corporate headquarters, spin-offs, and increased local hiring, as well as the spread of telework and remote meetings brought about by the COVID-19 pandemic. If these changes are accompanied by changes in values related to residential preferences, there is a possibility that they will turn into a major trend countering the over-concentration in the Tokyo area. From the viewpoint of regional population distribution, we must continue to keep a close eye on the changes in the domestic migration trends due to the COVID-19 pandemic.

[7] (J): written in Japanese.

Koike S, Shimizu M (2020) *Tokyo-ken ikkyoku shuchu ha keizoku suruka? Shusshochi bumpu henka kara no kensho* (Will population concentration continue in the Tokyo Area? An investigation into birthplace distribution). *Jinko Mondai Kenkyu* (Journal of Population Problems), 76(1): 80–97 (J)

Koike S, Suga K, Kamata K et al. (2020) *Nippon no chiiki-betsu shorai suikei jinko kara mita shorai no shusshosu* (Municipal birth projections consistent with IPSS (2018) regional population projections for Japan 2015–2045). *Jinko Mondai Kenkyu* (Journal of Population Problems), 76(1): 4–19 (J)

Nakagawa M, Yamauchi M, Suga K et al. (2018) *Todofuken-betsu ni mita gaikokujin no shizen dotai* (Prefecture-level natural changes of the foreign population in Japan). *Jinko Mondai Kenkyu* (Journal of Population Problems), 74(4): 293–319 (J)

Suga K, Koike S, Kamata K et al. (2020) *Nippon no chiiki-betsu shorai suikei jinko kara mita shorai no shibosu* (Municipal death projections consistent with IPSS (2018) regional population projections for Japan 2015–2045). *Jinko Mondai Kenkyu* (Journal of Population Problems), 76(1): 20–40 (J)

Chapter 4
Future Projection of Accessibility to Hospital Beds in the Suburbs: The Case of the Northern Osaka Metropolitan Area

Ryo Tanimoto

Abstract Lack of accessibility and differential access to various facilities and services in urban areas of Japan, especially access to medicine and health care, is expected to become problematic in the near future. The author examined such a likely transformation and the problems of accessibility to hospital beds in a suburban residential area (northern metropolitan Osaka) by analyzing accessibility using the two-step floating catchment area method and evaluating current urban planning. The analysis conducted in this study identified two different accessibility problems related to hospital beds: insufficient supply and disparity among transportation modes or neighborhoods. By 2025, the author's accessibility projections, based on improvements in public transportation and controls on the number and types of medical care beds, suggest that the actual resolution of such problems requires a multifaceted approach to using the existing resources in the study area. The results identified two requirements for improving accessibility to medical facilities: (1) the promotion of cooperation between planning implemented by each municipality and specialized medicine/health policies that target broader areas, beyond individual municipalities and (2) the continuation of a flexible and critical examination into the validity of current urban planning by each municipality.

Keywords Future projection · Accessibility · Hospital beds · Suburban residential areas · Two-step floating catchment area method

4.1 Introduction

Faced with an aging and declining population, Japan must make future projections and envision goals for specific regions by using population estimates and other means. In particular, future urban management must focus on the transformation of the suburbs of metropolitan areas, where a large-scale population decline is foreseen

R. Tanimoto (✉)
Faculty of Economics, Teikyo University, Hachioji-Shi, Tokyo, Japan
e-mail: tanimoto.ryou.dk@teikyo-u.ac.jp

© The Author(s), under exclusive license to Springer Nature Singapore Pte Ltd. 2023   51
Y. Ishikawa (ed.), *Japanese Population Geographies II*,
Population Studies of Japan, https://doi.org/10.1007/978-981-99-2076-1_4

(Miyazawa 2006). Making future projections of accessibility to a variety of goods and services, which are essential for improving quality of life and advancing social inclusion, is a valid approach to more accurately predicting the future of regions, including the actual conditions of people's lives (van Wee 2016; Pot et al. 2021; Tanimoto 2020).

Improving accessibility entails ensuring mobility for all people, as well as providing a stable supply of goods and services. Recently, however, the supply of medical and nursing care services in particular has reached a critical level even in major cities (Kawamata et al. 2008), leading to active policy discussions toward improving the situation. The 2012 amendment to the Long-Term Care Insurance Act stipulated the establishment of a "Community-based Integrated Care System" with the goal of providing medical and nursing care services within a 30-min distance from people's homes. In addition, the Act for Securing Comprehensive Medical and Long-Term Care in the Community, enacted in 2014, required the formulation of a "Regional Medical Care Vision" by each prefecture under the initiative of the Ministry of Health, Labour and Welfare. The Regional Medical Care Vision provides projections of the number of hospital beds that will be required in 2025, when the demand for medical care increases further as "baby boomers" (*dankai no sedai*, people born between 1947 and 1949) reach 75 years of age or older, and discusses measures to strengthen the medical care system based on these projections.

Consequently, accessibility to medical services in urban areas is expected to change significantly in the future. These projections are important for the accurate resolution of serious future regional problems. There are many previous and current studies on accessibility to health care. Examples of such studies overseas dealing particularly with urban healthcare facilities are those by Bell et al. (2013), which investigated primary healthcare provided by these facilities; by Loo and Lam (2012), which analyzed these facilities' relationship with the walking environment of the elderly; and by Mao and Nekorchuk (2013), which analyzed healthcare facilities in terms of transportation modes. Examples of studies on urban healthcare facilities in Japan include those by Hamasato (1999), Sekine (2003) and Kimura et al. (2011). The study of Takahashi (2015), which showed the national surplus in health care by region using the medical capacity of facilities and the temporal distance from homes, is also of interest. Some studies abroad have used the two-step floating catchment area method, which considers the balance between the population that can reach the facility and the capacity of the facility (Radke and Mu 2000; Luo and Qi 2009; Delamater 2013). In Japan, Masuyama (2015) used this method in a study of nursing care services. The above studies, however, have not specifically analyzed future changes in accessibility brought about by changes in population structure or by implementation of relevant policies.

Notable exceptions to this trend among previous studies are those by Akimoto (2014, 2016), which discussed the impact of the increased availability of public transportation and the consolidation of populations brought about by the implementation of compact city policies on future accessibility to facilities supporting daily life. However, in addition to the fact that these studies dealt with a somewhat special case of advanced cities implementing compact city policies, they did not address any other factor besides urban policies, such as the capacity of facilities and future changes.

In light of the above background, this study examines future issues of accessibility to hospital beds and offers predictive insights for their resolution. In particular, this study projects the accessibility to hospital beds for the population in 2025, with reference to that in 2010, in terms of scenarios involving the provision of beds and transportation systems. It also examines the effectiveness and limitations of the current urban and medical policies. The next section describes the analysis targets and methods. Section III elucidates the changes in accessibility to hospital beds from 2010 to 2025 and examines the differences in accessibility among different policy scenarios. Section IV gives the conclusions of the study and the remaining issues to address.

4.2   Definition of Targets and Methods of Analysis

1. Study area

This study covers the municipalities shown in Fig. 4.1 in a "sphere of daily life" comprising several built-up areas and new housing developments surrounding Senri New Town in the northern part of Osaka Prefecture. This area, located from around 10 to 30 km from the center of Osaka City, is a suburban residential area that spans Osaka Prefecture and Hyogo Prefecture. An extensive transportation network and dense population can be found south of Senri New Town (toward the metropolitan area), which has a population of about 90,000. Population clusters also exist in the northern region, despite the slightly less extensive public transport network and the higher aging rate. Senri Chuo, located in the center of the study area, is becoming an important hub for daily life among the residents of northern Osaka Prefecture, as more and more offices and stores have recently been built in the area (Ishikawa 2008). On the basis of the recent multinucleation of the urban suburbs and the increasing importance of suburban nucleations in the spatial structure of urban areas (Ishikawa 2008), the author believes that the sphere of daily life centered on the suburban nucleation in Senri New Town is a suitable study area.

The population in the study area was 2,165,676 in the 2010 Census, but Inoue (2016) estimated that it would decrease to 2,029,173 by 2025 and that the aging rate (percentage of the population over 65) would rise from 21.6 to 27.9% from 2010 to 2025. To build a sustainable city amid a declining and aging population, local governments are working to improve accessibility to health care by reviewing urban and medical policies. Since the population problem and its countermeasures

Fig. 4.1 Study area

are also concerns in other regions, the insights gained in this study can be expected to contribute to discussions on the living conditions in typical urban suburbs.

2. Facilities, hospital beds, patients and transportation modes analyzed in the study

Investigating the accessibility to hospital beds entails consideration of the cost of travel between user homes and facilities and the surplus or shortage of supply (capacity of facilities, such as the maximum number of hospital beds and patient capacity) against demand (number of patients). Because there are many different types of required hospital beds and transportation modes available to patients, there is a need to clarify which patient categories and hospital bed types would cause problems in the future. In this study, the ratio of the attainable supply of services within a certain temporal distance against the demand for hospital beds (number of inpatients) was defined as the "sufficiency rate," which was used to evaluate accessibility.

The following data were used to calculate the hospital bed sufficiency rate. First, the author surveyed the number of beds by type in each facility in the study area (including clinics) on the basis of responses to the 2014 Hospital Function Report (Osaka Prefectural Government Department of Health and Medicine 2014b; Hyogo Prefectural Government Department of Health and Welfare 2017[1]). The Ministry of

---

[1] For the small number of hospitals that had unanswered items in the Hospital Function Report and the hospitals that were closed or consolidated in the period after publication of the report in 2014 up to the time of writing, the number of beds by bed type was estimated in accordance with the information obtained from the websites of the hospitals and the medical institution disclosure websites of Osaka and Hyogo Prefectures, along with the report manual published on the website of the Ministry of Health, Labour and Welfare. For example, the beds for intensive patient management,

Health, Labour and Welfare (2015) defines the following four types of beds. The "advanced acute care bed" provides highly specialized medical care for patients in the acute stage, aimed at early stabilization of their condition. The "acute care bed" also provides medical care for patients in the acute stage but requiring more orthodox medical care, again aimed at early stabilization of their condition. The "convalescent care bed" provides medical and rehabilitation functions, aimed at patient discharge. The "chronic care bed" provides medical care for patients who need long-term medical treatment, severely disabled persons and those with intractable disease. Table 4.1 shows the number of hospitals and beds by bed type in the study area. The differences in the numbers of hospitals and beds in the 2010 and 2025 analyses are due to the fact that the 2025 analysis reflects the closure or construction of new hospitals between 2010 and the timeframe of that analysis.

Table 4.1 Basic data on hospitals, beds and patients in the study area

	Hospitals	Advanced acute care beds	Acute care beds	Convalescent care beds	Chronic care beds
Number of hospitals/beds					
2010	157	1,929	10,628	1,693	5,366
2025 (status quo)	156	1,986	10,484	1,843	5,366
2025 (reflecting plans by hospitals recorded in Hospital Function Report)	(156)	(2,299)	(9,991)	(2,082)	(5,307)
Projected percentage of patients in each age group					
0–14		46.3%	42.5%	5.0%	6.2%
15–59		15.3%	41.5%	29.1%	14.1%
60–74		14.0%	36.1%	35.5%	14.4%
75–		7.3%	28.8%	33.8%	30.2%
Projected number of patients					
2010		1,924	5,433	5,241	3,523
2025		2,035	6,239	6,337	4,725

---

such as those in intensive care units, were classified as advanced acute care beds, the convalescent rehabilitation beds were classified as convalescent care beds, the long-term care beds as chronic care beds, and the others as acute care beds.

Next, population data were used as the basis for projecting the number of patients needing hospital beds. In this study, to make an analysis at the small area (*chocho-aza*) level,[2] the author used the results of the 2010 Census for the 2010 population.[3] For the 2025 population, data from the "Web System of Small Area Population Projections for the Whole of Japan" were used (Inoue 2016).[4] The results of general population estimates of small areas using the survival rate method tend to be unstable because the population of each area is generally small. Inoue addressed this problem by geographically smoothing the demographic changes in a certain area over a certain period of time by using the population of the surrounding area and the distance from the surrounding area (Inoue 2017).

In this study, the number of patients was estimated by multiplying the hospitalization rate (ratio of the number of inpatients to the population) by the population. This method is in accordance with the estimation method used in the Regional Medical Care Vision described in section I (Ministry of Health, Labour and Welfare 2015). However, since the hospitalization rate by bed type was not disclosed, in this study, the rate was estimated using the following method. First, the number of inpatients by age group for all bed types in the analysis was calculated by multiplying the population in each age group by the hospitalization rate by age group for all bed types based on the patient survey conducted by the Ministry of Health, Labour and Welfare. Next, the ratio of inpatients by age group for each bed type to the total number of inpatients was calculated backward from the results of the estimation (using the hospitalization rate as of 2013) of the number of patients for each bed type for 2025 in the Osaka Prefecture Regional Medical Care Vision. Results of the calculations are given in Table 4.1, which are based on values only from Osaka Prefecture, since the Hyogo Prefecture Regional Medical Care Vision was not published at the time of the analysis. The projected number of patients for each bed type was obtained by multiplying the above percentage by the number of inpatients by age group for all bed types determined above and adding the values for all age groups. This is expressed in the following formula:

$$I_{if} = \sum_a (P_{ia} \times S_a \times V_{fa}), \tag{4.1}$$

where I_{if} is the projected number of patients for bed type f for small area i in the Census, P_{ia} is the population of age group a for small area i in the Census, S_a is the

---

[2] *Chocho-aza* refers to a small administrative and statistical unit in Japan that is roughly equivalent to a US census block group.

[3] The number of small areas used in the analysis was 2053.

[4] The difference between the 2025 population estimated by the National Institute of Population and Social Security Research and that estimated by the Web System of Small Area Population Projections for the Whole of Japan was around a few percentage points for the cities in the study area, but the estimate by the National Institute of Population and Social Security Research was approximately 10% smaller for Toyono Town. Since it was difficult to judge the superiority or inferiority of the estimation accuracy and because the analytical units were small areas, this study used the data from the "Web System of Small Area Population Projections for the Whole of Japan".

hospitalization rate for age group a against all bed types in Osaka Prefecture, and V_{fa} is the ratio of inpatients for bed type f against all inpatients for age class a in Osaka Prefecture.

The data used were for the following years. The populations analyzed in this study were for 2010 and 2025, while the patient survey, which is conducted every three years, was for 2014, the most recent survey conducted at the time of the analysis. The projected number of patients in the Regional Medical Care Vision was based on the hospitalization rate in 2013. Due to the nature of the data used, it was unavoidable to use data from different years. In particular, the 2010 population was estimated based on the hospitalization rate (S_a) in Osaka Prefecture in the 2011 patient survey, which is close to 2010, and the ratio (V_{fa}) of inpatients by age group for each bed type against the total number of inpatients calculated using the hospitalization rate data in Osaka Prefecture in 2013. Furthermore, for the 2025 population, the most recent value at the time of analysis, i.e., S_a, was based on the value from the 2014 patient survey, and V_{fa} was based on the same ratio of inpatients as that used for the 2010 population. Recently, due to the increase in over-the-counter medical expenses by the elderly and the efforts of medical institutions to eliminate unnecessary hospitalization, the hospitalization rate has nearly leveled off across the entire population. However, it is unrealistic to assume that such an increase in medical expenses and such efforts by medical institutions will continue in the future (Inami 2009).

In the analysis used for this study, it was assumed that patients in the study area would always use facilities within the study area, and the use of facilities by patients outside the study area was not considered. This is because it is estimated that around 70–80% of patients in the study area in 2025 will be hospitalized in the medical care zone where they live, and it is highly unlikely that there will be a one-way outflow of patients between the study area and the surrounding areas (Osaka Prefectural Government Department of Health and Medicine 2014a, b).

The projected numbers of patients by bed type in the study area based on the above assumptions are given in Table 4.1. Although the number of inpatients will increase overall, as shown in the table, since the percentage of patients in convalescent and chronic care beds in each age group is relatively large among the elderly, it is projected that the number of patients in those hospital beds will increase significantly by 2025, as the aging of the population advances.

Finally, the patient transportation modes were characterized as follows. In this study, in consideration of the necessity of examining accessibility to care based on various modes of transportation (Neutens 2015) and the guidelines of the Community-based Integrated Care System, which aims to avoid long-term hospitalizations as much as possible and encourage home care, two modes of transportation were assumed, namely by private car and by public transportation (railways, local buses and walking). The facilities available to the population of the small area were considered, for private cars, those that could be reached within 30 min from the place of residence (centroid of each small area) and, for public transportation, those that could be reached within 30 to 60 min. The travel time and frequency of operations between railway stations and bus stops were based on the timetable from each transportation operator for the daytime (10:00–15:00) on weekdays (as of the end of

June 2016).[5] It was assumed that toll roads were not used for private cars, and the road information was based on the 2016 edition of "ArcGIS Data Collection Road Network" published by ESRI Japan Corporation and Sumitomo Electric Industries, Ltd. The walking speed was set to 4 km/h.

3.   Method for calculating sufficiency rate

In this study, the sufficiency rate was calculated using the two-step floating catchment area method mentioned in the previous section. In particular, the following method by Wang and Luo (2005) was used.

Step 1: The sum of the number of patients for a given bed type at all residential areas (centroids of small areas) located within a certain distance d_0 (temporal distance in this study) is obtained from each facility, and the ratio of the number of beds for that type is calculated as

$$R_{jf} = \frac{B_{jf}}{\sum_{i \in \{d_{ij} \leq d_0\}} I_{if}}, \tag{4.2}$$

where R_{jf} is the ratio of the number of beds for bed type f at point j where the facility is located against the corresponding number of patients, i is the centroid of the district where the residents live (small area), B_{jf} is the number of beds for bed type f in facilities at point j, d_{ij} is the distance between i and j, and d_0 is a fixed distance.

Step 2: For each district (small area), all of the locations of facilities located within a fixed distance d_0 from the center are identified, and the sum of the R_{jf} for the identified facilities is obtained as shown in the following equation:

$$A_{if} = \sum_{i \in \{d_{ij} \leq d_0\}} R_{jf} = \sum_{i \in \{d_{ij} \leq d_0\}} \left(\frac{B_{jf}}{\sum_{c \in \{d_{cj} \leq d_0\}} I_{if}} \right), \tag{4.3}$$

where A_{if} is the number of available (accessible) beds per patient for bed type f in district i (i.e., accessibility based on the two-step floating catchment area method).

Equation (4.3) reveals the surplus or deficiency of the maximum number of available beds relative to the number of patients. That is, if A_{if} is greater than 1, the number of beds is greater than the number of patients, and vice versa if it is smaller than 1. However, the number of beds actually secured should include a margin in order to cope with short-term fluctuations in demand. The bed occupancy rates incorporating such margins were defined as follows with reference to data from the Ministry of Health, Labour and Welfare (2015): advanced acute care: 75%, acute care: 78%, convalescent care: 90%, chronic care: 92%. The degree of sufficiency of the number

---

[5] For railways, local trains that stop at each station were used as the standard, and waiting time was set to half of the average train interval. The waiting time for local buses was set to 10 min for sections with a frequency of more than one service every 20 min, 15 min for sections with a frequency of one service every 20 to 40 min and 20 min for sections with lower frequency of service.

of beds that factor in the bed occupancy rate relative to the number of patients for bed type f in district i, or, namely the sufficiency rate S_{if}, was defined as

$$S_{if} = A_{if} \times O_f, \qquad (4.4)$$

where O_f is the occupancy rate for bed type f. A sufficiency rate exceeding 100% means that the supply of hospital beds will remain sufficient even after factoring in short-term fluctuations in demand.

4.3 Future Changes in Accessibility to Hospital Beds

1. Current state analysis and future projection

This section discusses the analysis of accessibility to hospital beds for the populations in 2010 and 2025. First, the bed sufficiency rate was determined for the population in 2010, as well as for the population in 2025, by assuming that the number of beds and bed types at the time of analysis remains the same (maintenance of status quo) to clarify the changes and problems in the state of bed sufficiency in the near future.

The italicized values of Table 4.2 show the descriptive statistics of the sufficiency rates for three patterns in 2010 and 2025. Figure 4.2 maps the sufficiency rate in the 30-min zone for private cars in 2025 (maintenance of status quo for the number of beds and hospital bed types) by bed type.

First, the sufficiency rate for advanced acute care beds was clearly divided into regions with a somewhat high sufficiency rate and those with an extremely low sufficiency rate (Fig. 4.2). This reflects the small number of facilities providing advanced acute care beds. Convalescent care beds also have an extremely low sufficiency rate due to the low number of beds provided. On the other hand, for acute care beds and chronic care beds, the sufficiency rate was high in the central and southern parts of the study area, where there are many facilities. A certain level of sufficiency is also ensured even in the northern part of the study area. Looking at the yearly changes from Table 4.2, we see that the average sufficiency rate decreases for all hospital bed types except for advanced acute care in the 30-min zone for public transportation. The decline in chronic care beds, where the number of patients increases significantly, was particularly large. The rate of decline in convalescent care beds was already extremely low as of 2010, and a considerable shortage of convalescent care beds persists.

By mode of transportation, the average sufficiency rate for the 30-min private car zone was generally higher than that for the 30-min public transportation zone. Also, the coefficient of variation[6] was generally larger for public transportation than for private cars. Figure 4.3 shows the sufficiency rate for the 30-min public transportation zone divided by the sufficiency rate for the 30-min private car zone in each small

---

[6] Statistic obtained by dividing the standard deviation by the mean; in the analysis, it was considered an indicator of the disparity between small areas.

Table 4.2 Descriptive statistics of sufficiency rates

	Car (30-min zone)			PT (30-min zone)			PT in scenario A (30-min zone)			PT in scenario B (30-min zone)		
2010	Av.	Med.	CV	Av.	Med.	CV	Av.	Med.	CV	Av.	Med.	CV
Advanced acute care	71.0	107.6	0.67	17.3	1.3	2.87						
Acute care	146.8	150.9	0.33	130.0	112.2	0.77						
Convalescent care	23.2	24.6	0.41	23.5	10.0	1.90						
Chronic care	115.9	110.5	0.34	108.8	90.5	0.78						
(status quo)												
Advanced acute care	66.2	78.7	0.68	23.1	1.2	2.38	39.9	1.2	2.54	44.8	6.2	1.90
Acute care	126.6	129.0	0.33	114.0	104.0	0.76	134.4	122.1	0.74	119.9	112.6	0.59
Convalescent care	21.1	23.3	0.35	22.8	13.9	1.59	25.9	14.6	1.46	21.8	13.4	1.16
Chronic care	86.9	83.2	0.34	83.3	72.8	0.78	93.7	75.8	0.86	84.3	78.3	0.59
Scenario I												
Acute care	90.6	91.1	0.33	82.5	72.0	0.78	96.6	83.8	0.76	86.2	79.4	0.60
Convalescent care	53.6	54.2	0.31	53.1	42.4	0.90	60.8	48.3	0.85	53.0	46.5	0.66
Scenario II												

(continued)

Table 4.2 (continued)

	Car (30-min zone)			PT (30-min zone)			PT in scenario A (30-min zone)			PT in scenario B (30-min zone)		
	Av.	Med.	CV	Av.	Med.	CV	Av.	Med.	CV	Av.	Med.	CV
Acute care	89.5	90.3	0.33	85.5	80.4	0.70	97.4	88.7	0.70	87.5	82.8	0.55
Convalescent care	54.7	53.5	0.30	50.3	42.5	1.02	60.1	45.7	0.94	51.7	47.6	0.74

	PT (60-min zone)			PT in scenario A (60-min zone)			PT in scenario B (60-min zone)		
	Av.	Med.	CV	Av.	Med.	CV	Av.	Med.	CV
2010									
Advanced acute care	53.3	12.6	0.94						
Acute care	138.7	147.9	0.36						
Convalescent care	22.1	16.9	0.58						
Chronic care	107.4	105.3	0.34						
2025 (status quo)									
Advanced acute care	52.5	44.9	0.88	60.2	54.2	0.85	64.8	73.3	0.70
Acute care	117.7	127.8	0.34	127.5	133.2	0.34	124.7	123.3	0.29
Convalescent care	20.5	19.0	0.47	20.9	19.6	0.45	20.8	22.0	0.31
Chronic care	80.7	78.7	0.34	83.5	81.9	0.34	83.1	86.5	0.28
2025 Scenario I									
Acute care	84.6	92.3	0.34	91.4	95.1	0.34	89.4	87.2	0.30
Convalescent care	51.1	53.4	0.34	53.7	56.9	0.33	52.9	53.8	0.26
2025 Scenario II									
Acute care	84.7	90.0	0.34	90.0	94.5	0.33	88.7	85.9	0.29
Convalescent care	51.0	54.6	0.36	55.1	60.1	0.34	53.6	53.8	0.27

CV Coefficient of variation

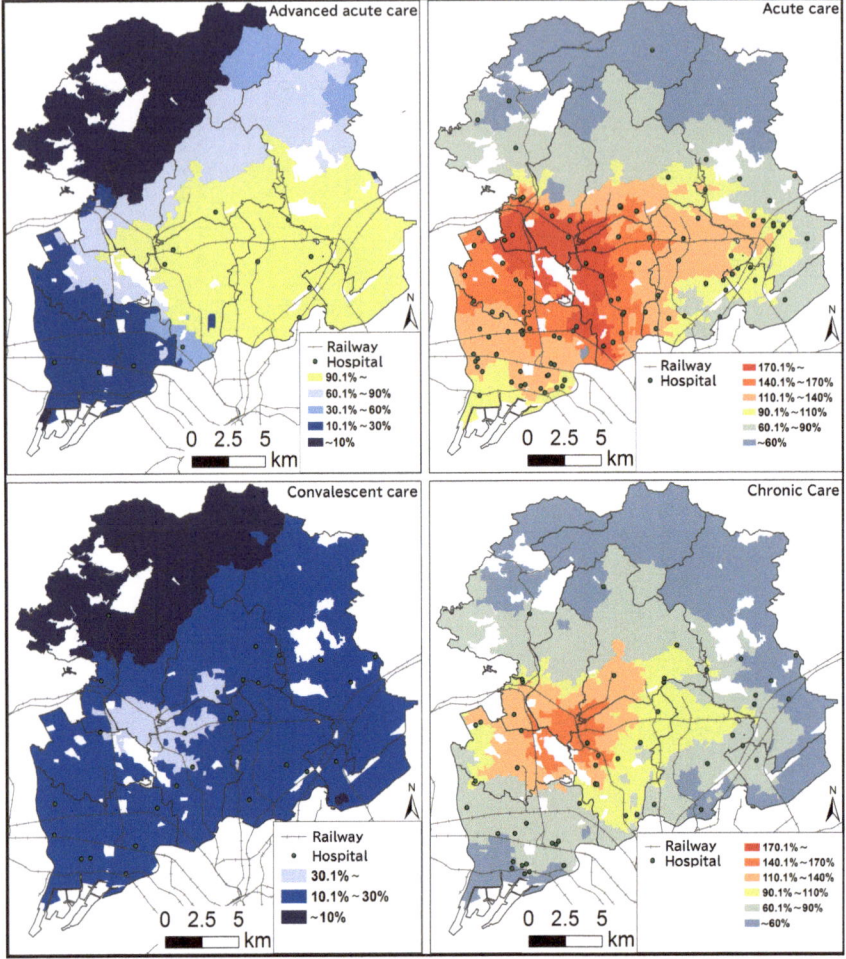

Fig. 4.2 Sufficiency rate for beds in terms of 2025 population (30-min zone by car)

area for acute care beds. The sufficiency rate for public transportation was higher than that for private cars only in the vicinity of railway stations. This implies that areas that are far from stations and are not conveniently accessible by local buses tend to be at a disadvantage in terms of sufficiency rate.

Looking at the geographical characteristics of the sufficiency rate in light of these results, we can see that the sufficiency rate is low for all bed types in most of the districts in the northern part of the study area where there are few facilities. Since there are few railway lines in this area, and there are many sections where buses operate less frequently, the disadvantage in sufficiency rate when there is no access to private cars is likely to become more pronounced. However, even in the central and southern regions where there are many facilities, areas with low sufficiency rates

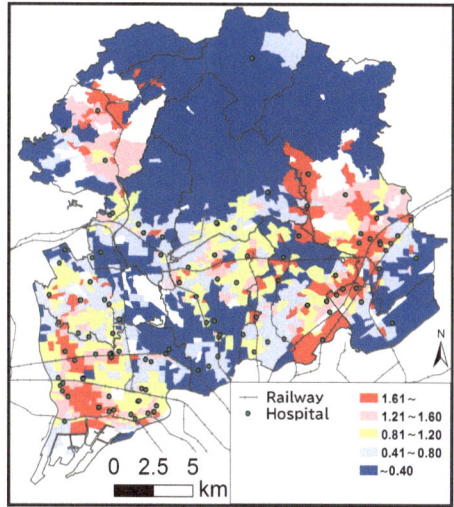

Fig. 4.3 Comparison of sufficiency rate for acute care beds in terms of 2025 population between 30-min zones by car and by public transportation

occur in a mosaic pattern. The lack of guaranteed access to a sufficient number of hospital beds, even in areas where there are many hospitals, is seen as problematic.

In summary, the issue pertaining to accessibility to hospital beds now and in the future can be summed up as "shortage and disparity." First, "shortage" refers to the situation in which there is an absolute shortage of all hospital bed types except for acute care, which will further worsen in the near future. In other words, the analysis showed that the level of sufficiency is not stable even with the use of private cars and even in the central to southern regions of the study area where there are many hospitals. "Disparity," on the other hand, refers to the relative disadvantage in sufficiency rate due to the available modes of transportation and the location of the districts. In terms of the sufficiency rate, public transportation is disadvantageous compared with private cars, especially in areas where there are no railway lines and residents rely on local buses. Regions such as the northern part of the study area, where a certain number of people live despite the inconvenience of transportation, must deal with both "shortage and disparity."

Given the prolonged recession and population decline, it is more realistic to make up for the shortage of medical resources by utilizing other existing medical resources than to construct a large number of new facilities. For example, there was a certain margin in the number of acute care beds in the study area relative to the number of patients even for 2025 (Table 4.1), and the sufficiency rate was relatively high (The italicized values of Table 4.2). This surplus can be diverted into other hospital bed types. The Hospital Function Report in fact indicates whether each hospital has plans to change its bed types by 2025, as well as the number of beds to be changed for each bed type. Few hospitals, however, plan to make large changes in hospital bed types (from Table 4.1: number of hospitals/beds: 2025 (reflecting plans by hospitals recorded in Hospital Function Report)), making it unlikely that these changes alone would solve the "shortage and disparity" in hospital beds. In particular, the "shortage and disparity" in convalescent care beds will become a serious problem in the near future as the population ages further.

2. Policy scenario analysis

Next, in order to alleviate the problem of "shortage and disparity" in convalescent care beds and the disadvantage of public transportation, the author conducted an analysis based on scenarios in which different measures are taken in terms of policies on hospital bed type and transportation. First, in regard to hospital bed types, it was assumed that approximately 25% of acute care beds were additionally converted to convalescent care beds from the "number of hospitals/beds in 2025 (reflecting plans by hospitals recorded in Hospital Function Report)" in Table 4.1. A 25% conversion rate was deemed acceptable in consideration of the short-term fluctuations in demand for beds and the approximately 30% margin in the number of acute care beds relative to the projected number of inpatients in the study area for 2025 (Table 4.1). Even if the number of beds to be converted is the same, if the number of hospitals involved in the conversion of beds is different, a spatial difference in sufficiency rate would likely occur. Two scenarios, therefore, were assumed, namely Scenario I, where all hospitals with acute care beds converted 25% of their total number of beds, and Scenario II, where only hospitals of a certain size (hospitals with 200 or more acute care beds) converted 50% of their beds to convalescent care beds[7]. Scenario I assumed changes in hospital bed types in 104 facilities and a supply of 4572 beds from a total of 119 facilities, including existing facilities. Scenario II, on the other hand, assumed changes in bed types in 13 facilities and a supply of 4658 beds from 42 facilities.

Two scenarios were also assumed for public transportation: Scenario A assumed the completion of the extension of railway lines and the opening of new stations by 2025 (planned new railway stations in Fig. 4.1). Scenario B assumed the shortening of bus waiting times to no more than 10 min, in addition to the extension of railway lines and the opening of new stations in Scenario A.

For Scenarios I and II, related to hospital bed types, the scenarios for the 30-min private car zone and the maintenance of the status quo for public transportation, as discussed in the previous section, were also analyzed. For Scenarios A and B, related to public transportation, the scenario for the maintenance of the status quo for hospital bed types was also analyzed. Thus, cases where measures were implemented either only for hospital bed types or only public transportation were also examined. Crossing the settings for hospital bed types as of 2025 (three types: status quo and Scenarios I and II) with the settings for transportation modes (seven types: 30-min zone for private cars, 30- and 60-min zones for the status quo for public transportation and 30- and 60-min zones for Scenarios A and B) resulted in 21 scenario patterns. The descriptive statistics of 18 of these patterns (excluding the three patterns for 2025

---

[7] The reason why we set the conversion of acute care beds to 50% in hospitals with more than 200 acute care beds in Scenario II was to keep the number of beds to be converted at the same level as Scenario I, assuming that the hospital beds will be converted based on clear policy criteria. The hospital bed number of 200 is a traditional criterion used in determining hospital scale, such as in the classification of medical fees.

mentioned in the previous section that are not related to Scenarios I or II and Scenarios A or B) are shown in the non-emphasized parts of Table 4.2. The suggested effects of these measures on the hospital bed sufficiency rate are described as follows.

First, in regard to hospitable beds, when a large number of the surplus acute care beds were converted to convalescent care beds, the average sufficiency rate for convalescent care beds was significantly improved without significantly reducing the sufficiency rate for acute care beds throughout the study area. Scenario I, where a large number of facilities implemented conversion, showed smaller regional disparity (coefficient of variation) in convalescent care beds than did Scenario II.

With regard to public transportation, in Scenario A, the average sufficiency rate for all hospital bed types was significantly higher than when the status quo for public transportation was maintained. However, there was no significant change in the coefficient of variation, or in regional disparity. This suggests that expanding the railway network only in some districts provides benefits that are limited to the areas near the railway stations in those districts, and it has minimal effect in improving convenience in areas that rely on terminal transport where there is a low sufficiency rate to begin with due to the unavailability of public transportation. On the other hand, the coefficient of variation decreased significantly in Scenario B. Figure 4.4 shows that there are areas with high sufficiency rates for convalescent care beds even outside those near railway stations in Scenarios I and B (30-min zone). This shows that improving the convenience of local buses significantly alleviates regional disparity. This tendency is particularly notable in the short time range (30-min zone).

However, in the analysis for all of the scenarios, the sufficiency rate for convalescent care beds remained below 100%. Therefore, the conversion of surplus acute care beds alone cannot completely compensate for the shortage of convalescent care beds. Further increasing the sufficiency rate involves increasing the total number of convalescent care beds. It has become clear, however, that a combination of improving

Fig. 4.4 Sufficiency rate for convalescent care beds in terms of 2025 population (scenario of bed functions I and PT B)

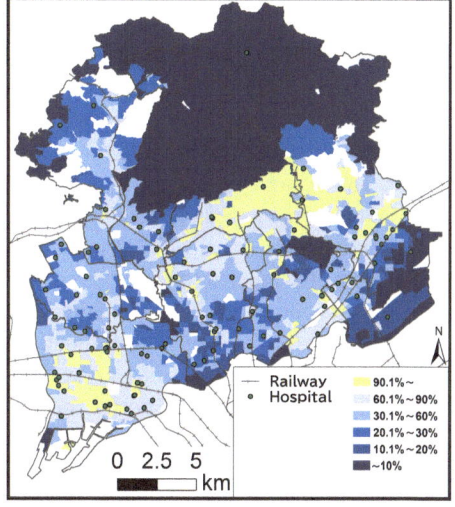

public transportation and converting hospital bed types in the study area alleviates "shortage and disparity" to a certain extent. The results also suggest that for the same number of hospital beds to be provided, distributing their provision among a larger number of facilities is more advantageous in alleviating regional disparities.

4.4   Conclusion

The purpose of this paper was to examine the future issues of accessibility to hospital beds in the northern part of the Osaka Metropolitan Area and to offer predictive insights for their resolution. Mainly, this study showed that the practical alleviation of the "shortage and disparity" in accessibility to hospital beds, which is expected to further worsen in the future, requires a multifaceted approach, such as the adjustment of hospital bed types, the improvement of transportation modes and the reconsideration of facility locations. The results of this study suggest that, implemented singly, the improvement of public transportation and the adjustment of hospital bed types each has only a limited effect. Combined, however, they can improve accessibility to a certain extent without a major increase in the number of hospital beds.

In order to coordinate measures implemented by different agencies and departments, such as urban and medical policies, it is necessary to solve the problem of bureaucratic sectionalism. Improving accessibility to health care requires the creation of a mechanism for local governments to verify the validity of urban planning on the basis of technical materials on the actual conditions of facilities and services, as well as data on the behavior of users across a wide area, including neighboring municipalities, and on the sufficiency of services. For example, coordination between legal systems and departments must be enhanced. Results and publications of prefectural medical and health departments, such as the Regional Medical Care Vision and the Hospital Function Report, must be put to wider use in the evaluation of urban plans and of the current situation of the regions by the urban policy departments of cities. Such coordination of policies on different spatial scales, such as urban planning by each municipality and specialized medical and health policies that target broader areas, will be important in the future for regions oriented toward compact city planning. Each municipality must also continue to examine the validity of its urban plans flexibly and critically.

Although this paper specifically examined the case of a suburb of a major metropolitan area, fiscal difficulties and the aging and decline of the population are problems for Japan as a whole. Predicting the future, therefore, in terms of accessibility to resources is crucial in addressing the future regional problems brought about by these social issues. Specifically, it is important to predict future regional disparities in accessibility in order to address various problems by appropriately allocating the limited human, material and economic resources. Other than presenting insights obtained from a specific regional case, the author believes that this paper's work will also contribute to the study of methodologies for examining future regional disparities in accessibility. In particular, the accessibility projection method used in this paper, setting aside the data constraints, can be applied to other regions and services.

Finally, the following are some limitations of this study. First, the discussion in this paper is, undeniably, severely restricted by the conditions set in the analysis and by the available data. Moreover, the enhancement of facilities and services must be discussed by taking construction and personnel costs into account. In addition to the quantitative accessibility to health care discussed in this paper, accessibility to various financial and activity opportunities, as well as perceived accessibility, which is measured in terms of how individuals feel about the ease of access, also affects quality of life. There is a need, therefore, to comprehensively measure the quality of life while reflecting the diversity of residents and their environmental conditions. The author has previously delved into discussions of the relationship between accessibility and quality of life through the creation of indices for the overall sense of accessibility to daily life necessities (Tanimoto and Hanibuchi 2021). Future projections based on this relationship should also be made going forward.

References[8]

Akimoto N (2014) Toyama-shi no kurasuta gata kompakuto shitei seisaku to kougai no akuse-shibiritei: Fuchu chiiki ni okeru shimyureshon (Implications of a cluster-type compact city design for accessibility in the suburbs of Toyama City: Simulations in the Fuchu area). Chirigaku Hyoron (Geogr Rev Jpn Ser A) 87: 314–327. https://doi.org/10.4157/grj.87.314 (J)

Akimoto N (2016) Ikkyoku shuchu gata kompakuto shitei seisaku no yukosei ni kansuru akuse-shibiritei no shimyureshon bunseki: Aomori-shi ni okeru jinko no shuyakuka to kokyokotsu no tahindoka (Implications of centralized compact city designs for accessibility from simulation analyses: Improving population distribution and public transportation in Aomori City). Chigaku Zasshi (J Geogr) 125: 523–544. https://doi.org/10.5026/jgeography.125.523 (J)

Bell S, Wilson K, Bissonnette L et al (2013) Access to primary health care: Does neighborhood of residence matter? Ann Assoc Am Geogr 103: 85–105. https://doi.org/10.1080/00045608.2012.685050

Delamater PL (2013) Spatial accessibility in suboptimally configured health care systems: A modi-fied two–step floating catchment area (M2SFCA) metric. Health Place 24: 30–43. https://doi.org/10.1016/j.healthplace.2013.07.012

Hamasato M (1999) Iryokikai eno akuseshibiritei kara mita Okinawa-honto chiiki no kukankozo (The spatial structure in Okinawa Island in terms of accessibility to medical opportunities). GIS Riron to Oyo (Theory and Applications of GIS) 7: 35–42. https://doi.org/10.5638/thagis.7.2_35 (J)

Hyogo Prefectural Government Department of Health and Welfare (2017) Byosho kino hokoku heisei 26 nendo (Hospital function report in 2014) https://web.pref.hyogo.lg.jp/kf15/byousyouk inou.html. Accessed 3 Mar 2017 (J)

Inami I (2009) Shakaiteki nyuin no kenkyu: koureisha iryou saidai no byori ni ikani taisho subekika ("The study on "social hospitalization": How should we cope with the most difficult problem in the medical care for the elderly?). Toyo Keizai Inc., Tokyo (J)

Inoue T (2016) Zenkoku shochiiki betsu shoraijinko suikei shisutemu (The web system of small area population projections for the whole Japan). http://arcg.is/1LqC6qN. Accessed 30 Nov 2016 (J)

---

[8] (J): Written in Japanese.

Inoue T (2017) A new method for estimating small area demographics and its application to long-term population projection. In: Swanson DA (ed) The frontiers of applied demography. Springer, Heidelberg, p 473–489

Ishikawa Y (2008) Kogai kara mita toshiken kukan; Kogaika takakuka no yukue (Metropolitan area space from a suburban perspective: The future of suburbanization and polycentralization). Kaiseisha, Otsu (J)

Kawamata K, Hirota E, Honda T et al (2008) Chiiki iryo hokai no kiki: Shutoken demo!? (Crisis of community medicine: Even in the Greater Tokyo Area?) Honnoizumisha, Tokyo (J)

Kimura Y, Hamano T, Shiwaku K (2011) Chiri joho sisutemu (Geographic Information System; GIS) o mochiita Shimane-Ken ni okeru kyukyuhanso kabaritsu ni kansuru kento (Evaluation of coverage for emergence medical services in Shimane Prefecture using geographic information system) Nihon noson igakukai zasshi (J Jpn Assoc Rural Med) 60: 66–75. https://doi.org/10.2185/jjrm.60.66 (J)

Loo BPY, Lam WWY (2012) Geographic accessibility around health care facilities for elderly residents in Hong Kong: a microscale walkability assessment. Environ Plann B 39: 629–646. https://doi.org/10.1068/b36146

Luo W, Qi Y (2009) An enhanced two-step floating catchment area (E2SFCA) method for measuring spatial accessibility to primary care physicians. Health Place 15: 1100–1107. https://doi.org/10.1016/j.healthplace.2009.06.002

Mao L, Nekorchuk D (2013) Measuring spatial accessibility to healthcare for populations with multiple transportation modes. Health Place 24: 115–122. https://doi.org/10.1016/j.healthplace.2013.08.008

Masuyama A (2015) Aomori-ken Hirosaki-Shi ni okeru kaigo sabisu eno akuseshibiritei keisoku no kokoromi (Measuring the spatial accessibility to home care services: A case study of Hirosaki-shi, Aomori). Toshikeikaku Ronbunshu (J City Plann Inst Jpn) 50: 210–220. https://doi.org/10.11361/journalcpij.50.210 (J)

Ministry of Health, Labour and Welfare (2015) Chiiki iryo koso sakutei gaidorain (Guideline for planning regional medical care vision). http://www.mhlw.go.jp/file/06-Seisakujouhou-108 00000-Iseikyoku/0000088510.pdf. Accessed 30 Nov 2016 (J)

Miyazawa H (2006) Katoki ni aru daitoshiken no kogai nyutaun: Tama nyutaun o jirei ni (The changing face of suburban new towns in large metropolitan areas: The case of Tama New Town, Tokyo). Keizaichirigaku Nempo (Ann Assoc Econ Geogr) 52: 236–250. https://doi.org/10.20592/jaeg.52.4_236 (J)

Neutens T (2015) Accessibility, equity and health care: review and research directions for transport geographers. J Transp Geogr 43: 14–27. https://doi.org/10.1016/j.jtrangeo.2014.12.006

Osaka Prefectural Government Department of Health and Medicine (2014a) Osaka-Fu chiiki iryo koso (Osaka Prefecture regional medical care vision). http://www.pref.osaka.lg.jp/attach/2502/00213231/zenpen.pdf. Accessed 30 Nov 2016 (J)

Osaka Prefectural Government Department of Health and Medicine (2014b) Heisei 26 nendo byosho kino hokoku (Hospital function report in 2014b). http://www.pref.osaka.lg.jp/iryo/keikaku/26b yousyoukinou.html. Accessed 3 Mar 2017 (J)

Pot FJ, Van Wee B, Tillema T (2021) Perceived accessibility: What it is and why it differs from calculated accessibility measures based on spatial data. J Transp Geogr 94: 103090. https://doi.org/10.1016/j.jtrangeo.2021.103090

Radke J, Mu L (2000) Spatial decompositions, modeling and mapping service regions to predict access to social programs. Geogr Inf Sci 6: 105–112. https://doi.org/10.1080/10824000009480538

Sekine T (2003) Kinsetsusei no jikukanteki anteido no bunseki: Chiba-ken Matsudo-shi no ganka iin o jirei ni shite (Analysis of spatio-temporal stability in accessibility: A case study of ophthalmic hospitals in Matsudo City, Chiba Prefecture). Chirigaku Hyoron (Geogr Rev Jpn) 76: 725–742. https://doi.org/10.4157/grj.76.10_725 (J)

Takahashi T (2015) Zenkoku kakuchi no iryo/kaigo no yoryoku o hyokasuru (Assessing the medical and long-term care capacity of Japan). In: Masuda H (ed) Tokyo shometsu: Kaigohatan to chihoiju (Tokyo's disappearance: Nursing care failure and rural migration) Chuokoron-shinsha, Tokyo, p 77–95 (J)

Tanimoto R (2020) Seikatsu no shitsu ni kakawaru akuseshibiritei kenkyu no seika to kadai: 1980 nendai iko no doko o chushin ni (Achievements and issues of studies on accessibility relating to quality of life: Trends since the 1980s). Jimbun Chiri (Jpn J Hum Geogr) 72: 361–381. https://doi.org/10.4200/jjhg.72.04_361 (J)

Tanimoto R, Hanibuchi T (2021) Associations between the sense of accessibility, accessibility to specific destinations, and personal factors: A cross-sectional study in Sendai, Japan. Transp Res Interdiscip Perspect 12: 100491. https://doi.org/10.1016/j.trip.2021.100491

Van Wee B (2016) Accessible accessibility research challenges. J Transp Geogr 51: 9–16. https://doi.org/10.1016/j.jtrangeo.2015.10.018

Wang F, Luo, W (2005) Assessing spatial and nonspatial factors for healthcare access: towards an integrated approach to defining health professional shortage areas. Health Place 11: 131–146. https://doi.org/10.1016/j.healthplace.2004.02.003

Chapter 5
Toward a Politico-Economic Population Geography: A Critique of *The Shock of a Shrinking Japan*

Takashi Nakazawa

Abstract A series of population problems, such as a low fertility rate and population aging, has become the most urgent issue for Japanese society. These problems are tightly intertwined with the uneven distribution of population throughout the country. Tokyo's over-centralization has intensified even as a considerable number of settlements have neared the edge of extinction. Essentially, the most important policy target of contemporary Japan involves population geography. While population geographers have certainly been contributing to the practice of policymaking by illustrating the current population conditions, they have distanced themselves from disputes on the ideas or ideologies underlying population policy. In reviewing a timely published book titled *Shukusho Nippon-no Shogeki* (*The Shock of a Shrinking Japan*), this chapter explores the possibility of establishing a politico-economic population geography. Finally, the chapter concludes with the insistence that the concept of population itself, usually assumed to be a value-free geographical quantity, be reconsidered.

Keywords Geopolitics of population · Population decline · Regional revitalization · Geographic information · Othering

5.1 Introduction

In May 2014, the Japan Policy Council, a private organization, issued a report on its policy recommendations. The report designated 896 municipalities (49.8% of all municipalities) as "cities at risk of extinction"; in these municipalities, the population of women aged 20 to 39 years is projected to decrease to less than 50% by 2040 (Japan Policy Council, Subcommittee on Depopulation Issues 2014). Since the outflow of young women is also an outflow of fertility, for local governments, this phenomenon means a decline in their future populations. At the same time, the Tokyo metropolitan area continues to attract young people from rural areas. However, due to this city's

T. Nakazawa (✉)
School of Business Administration, Meiji University, Tokyo, Japan
e-mail: nkzw23@meiji.ac.jp

© The Author(s), under exclusive license to Springer Nature Singapore Pte Ltd. 2023   71
Y. Ishikawa (ed.), *Japanese Population Geographies II*,
Population Studies of Japan, https://doi.org/10.1007/978-981-99-2076-1_5

extremely low fertility rate, the concentration of women of childbearing age in Tokyo is bound to further accelerate Japan's declining birthrate. Therefore, the Japan Policy Council argued that it is necessary to halt the Tokyo metropolitan area's monopoly on net migration and to relocate the young people who had moved to the metropolitan area back to the regional areas. In September 2014, only four months after the publication of the report, the government established the Headquarters for Overcoming Population Decline and Vitalizing Local Economy in Japan. This prosaically named organization was positioned as the command center for dealing with depopulation, which had already become a reality, and addressing the sluggish regional economies (Headquarters for Overcoming Population Decline and Vitalizing Local Economy in Japan 2015).

The recognition that the declining population and the over-concentration in Tokyo are grave issues gradually became widespread throughout Japanese society via the mass media, as well as a best-seller paperback (Masuda 2014), and this recognition was actually elaborated by the Japan Policy Council. Thus, when NHK aired the program *Shukusho Nippon-no Shougeki* (*The Shock of a Shrinking Japan*) in September 2016, it elicited a tremendous response. While the program focused on raising awareness of the declining population as a crisis highlighted by the Japan Policy Council and Headquarters for Overcoming Population Decline and Vitalizing Local Economy in Japan, it also realistically depicted the geographically distinctive effects of the declining populations in the regions and the differing responses of residents and local governments. The contents of the program were later published as a book under the same title (NHK Special Crew 2017). This chapter explores the possibility of establishing a politico-economic population geography by examining how various actors in various regions are trying to deal with the problem of population decline and the over-concentration in Tokyo and how the authors of the book presented these issues.

Japan's problems, such as the over-concentration in Tokyo, the outflow of population from the regions, and the declining birthrate and aging population, occur in the social domain where population and geography are inseparably intertwined. The methodologies developed over many years in population geography can still be used for analyzing the current situation in regard to these issues. However, the traditional methodologies in population geography are not useful for analyzing the politics and the ideas behind the attempts to control the natural and social dynamics of the population toward achieving the "ideal state." Therefore, it is necessary to explore the possibility of establishing a population geography beyond quantitative analysis—one toward qualitative, political, and economic analysis.

The goal of this paper is consistent with those of the "geopolitics of population" (Bailey 2005) and "political demography" (Robbins and Smith 2017). Since the 1980s, population geography, like other fields of human geography, has been confronted by criticisms against positivism and empiricism. According to Bailey (2005), conservative population geographers held on to their strong identity as demographers, while the eclectics tried to find a compromise to deal with these criticisms by adopting qualitative methods in addition to quantitative methods. Meanwhile, the most radical researchers questioned the politics surrounding the concept of population itself by criticizing the essentialism lurking behind the concepts of race and ethnicity and rejecting reductionism, which claims that the whole is the sum of the parts. The "geopolitics of population" lies within the radical school of thought and refers to the geographic strategy employed by states and their agents over population groups in order to re-establish their authority (Bailey 2005).

The following is a very brief summary of the structure and contents of the book, *The Shock of a Shrinking Japan*. The prologue presents a basic recognition of the problems that will be caused by the rapid population decline and consequent expansion of uninhabited areas in various regions of Japan. In Chap. 1, "The future of over-concentration affecting Tokyo," citing the case of Toshima Ward, which has been recognized as a "city at risk of extinction" by the Japan Policy Council, the authors point out that the formation of families has not progressed, although many young people have moved to Toshima Ward; furthermore, this ward's existence as a municipality has been threatened by the low fertility rate and increased social security costs. Chapters 2 and 3, "Retreat strategy of a bankrupt city (1) and (2)," deal with the case of Yubari City, a run-down former coal mining town in Hokkaido. The authors depict the struggles of the local government trying to cut public services to the bone in the face of an extremely tight budget due to bankruptcy, as well as those of the residents who cannot leave the city and have to cope with the consequences of the government's actions. In Chapter 4, "Lack of access to even basic public services," the authors use the case of a certain district in Unnan City, Shimane Prefecture, to discuss the activities of a local resident organization, which has attracted attention since it became more difficult to maintain the standard of public services due to population decline. In this chapter, "Collapse of the local community: a disappearing village," the authors point out the serious problem faced by many local resident organizations, where aging residents must support the future of the region with a limited budget. In the epilogue, "'Over-concentration of death' that has begun in the suburbs of Tokyo," the authors go back to the Tokyo metropolitan area to portray the future awaiting those who are growing old alone, as told through the story of the increasing number of people dying without leaving anybody to look after their graves in Yokosuka City, Kanagawa Prefecture.

5.2 Use of Maps and Geographic Information

Maps are drawn as a means of exerting power and are always political in nature (Black 1998). On the other hand, maps also help empower the vulnerable primarily because of their political nature (Seager 2003). Maps appear throughout *The Shock of a Shrinking Japan*. In particular, in Chaps. 2 and 4, maps and geographic information are used for contrasting purposes, providing good material for a critical examination of the duality of maps and geographic information.

Chapter 2 presents the case of Yubari City, which was forced to implement extreme fiscal tightening. Due to bankruptcy resulting from the collieries' closure, the population of Yubari has fallen to less than one-tenth of its peak population. Public services must be cut back to a corresponding scale under a very tight budget. However, since physical infrastructure could not be consolidated flexibly in accordance with the declining population, the per capita administrative costs increased as the population decreased. City authorities, therefore, decided to introduce a system that visualizes the administrative costs of each region using a map and presented this system to the residents, and according to the book, "This makes it possible to visualize details like 'the cost of running a water pipe for only one house in this area is 0.2 million yen per year'" (p. 68). In other words, the local government used a map to "identify areas where the population is declining and administrative costs are inefficient as a way to guide people away from living in those areas" (pp. 68–69).

The local government mapped the geographic information that validates and authorizes its policy orientation, and the residents passively accepted its message. While this contributed to the backward-looking motivation of optimizing the retreat strategy, it did not motivate the residents to proactively come up with ideas to improve their own lives. In this context, maps and geographic information contributed to the survival of an administrative body, i.e., Yubari City, as "agents of persuasion" (Peck 2002) to urge residents to relocate.

While the residents of Yubari were persuaded by the authority of the map produced by the local government, the residents of the Nabeyama district in Unnan City, which is featured in Chap. 4, were proactive map makers. Nabeyama was confronted with a projection that its population would be halved in 20 years. Hirokazu Sakuno, a geographer at Shimane University, spoke directly to the community organization's president, who was skeptical about the projection. Using a handwritten map of the village on a whiteboard, he suggested that the residents themselves should choose which areas of the village to preserve and utilize, and which to return to forest, in order to consolidate the spaces in the village. "If you prepare now, even if it's hard, you will be able to preserve the village as a place where you will truly want to live until the end" (p. 148), said Sakuno to encourage the residents.

At the suggestion of Sakuno, the president of the community organization walked around the village again with a map in hand and realized that the land was turning into wasteland more widely than he had thought. After this, the residents decided to visualize the situation of the entire district using a map, and on its basis, they discussed the future of the village, including the consolidation of land use. Therefore, although a map was also used in Nabeyama district for the retreat strategy, it served as a base map for empowerment that brought an awareness of the various possibilities and directions of the retreat strategy. This is in contrast to the map used in Yubari City, which served as a route map to chart the retreat strategy along the government's existing policy directions.

These two contrasting uses of maps and geographic information are related to the geographic information system (GIS) debate and discussions on participatory GIS. In the early days of the GIS debate, anti-GIS advocates criticized it as being a tool to support crude positivism, reductionism, and utilitarianism and not to generate new knowledge (Taylor 1990; Taylor and Overton 1991; Sheppard 1993). The use of GIS in Yubari City could be subject to such criticisms, since it did not produce new geographical knowledge but simply functioned as an authority to guide the residents.

In the 2000s, as more and more researchers delved into the relationship between GIS and society, discussions on participatory GIS expanded (Kwan 2002; Sieber 2006). In the Nabeyama district, geographic knowledge that brought new regional awareness was generated through the creation of a map with the participation of residents. This can be considered an example of participatory GIS, with a handwritten map as its output. As this shows, participatory GIS does not necessarily require sophisticated software or a comprehensive system.

The cases of Yubari City and Nabeyama district are also relevant to the discussion on smart cities. In general, a smart city refers to a city that integrates information and communication technologies (ICT) in all areas, is governed by big data derived from such integration, and whose economy is driven by the creativity and entrepreneurship of "smart" people (Kitchen 2014). Recently, many critical studies have been published in Europe and America on neoliberal ideas, optimism for technocracy, and the surveillance of society in relation to the smart city concept (Kitchen 2014; Datta 2015; Luque-Ayala and Marvin 2015). In response, McFarlane and Söderström (2017) presented a project that aimed to restructure the smart city concept into one that contributes to the realization of social justice. In the project, participatory GIS was utilized, and marginal residents, who tend to be excluded from technology innovations, created the data with the help of experts. This provided an opportunity for the residents to become aware of the problems inherent in their living environment and to improve their current situation.

In a typical smart city, ICT is used for "smart growth" of the urban economy. Although, conversely, ICT was used in Yubari for retrogressive reasons, as in smart cities, the main purpose of ICT was to monitor the efficient operation of urban infrastructures. While the project conducted by McFarlane and Söderström (2017) was a large-scale participatory GIS project, the endeavor of Nabeyama district was based on a primitive handwritten map. Notwithstanding that, they are similar in that residents proactively collected geographic information and used the information to discover and solve problems in their districts.

5.3 Political Nature of Data

With the advances in ICT, it has become increasingly easy to acquire and use big data. As governance based on big data, such as that in smart cities, has become more common, questions on the nature of data have been raised (Gitelman 2013; Kitchen 2013, 2014). Data are often considered neutral, objective, and free from ideology. However, since data are selectively measured and created within certain technical constraints, they are not independent of the people involved and the social context surrounding the processes of conceptualizing the collection, gathering, processing, and analysis of data. Data are needed in social science to understand the state of society and changes in it. Furthermore, data cannot be separated from normative questions, such as how distant the actual is from the ideal and how to bring them closer. The recording of birth, death, and migration statistics by a state is closely related to the compulsion to grasp the current situation as a way to control the population and achieve the ideal state. Therefore, inevitably, "population research is political research" (Robins and Smith 2017: 212).

When such value-laden data are handled, processed, and presented in the form of maps and charts, the output becomes less and less objective and neutral. This is because the handling and processing of data are always done for a certain purpose, and the results are always presented in a specific context. For example, the population trend in Japan from 800 to 2100 A.D. is shown on pp. 4–5 of *The Shock of a Shrinking Japan*. It reveals that Japan's population has changed like a rollercoaster, with an explosive surge and then a drop after a gradual increase for more than a thousand years. According to the book, this trend is "not an expression of extreme pessimism but represents the envisioned future of the country" (p. 5) based on population statistics measured "objectively" and demographic estimates made through "scientific" procedures. However, viewed in the context conveyed by a title like *The Shock of a Shrinking Japan*, this trend can be seen as nothing but an embodiment of pessimism.

In fact, the conclusion of the book itself is pessimistic. In the epilogue, the authors state that they cannot offer a "prescription for the shrinking Japan" and conclude that the people of Japan "all have to throw themselves into a 'retreat strategy' while sharing the pain" (p. 196). The prescription they seek is one that would restore the dysfunction of the system that has supported Japan's economic growth in the past, in which the regional areas provide an abundant labor force for the Tokyo metropolitan area, and, at the same time, the wealth generated in Tokyo is redistributed to the regional areas in return for the labor force. No consideration is given to the idea of an alternative system that is not premised on this regional structure of labor supply and redistribution or to the argument that emphasizes ontological happiness backed by "spiritual affluence" rather than economic well-being based in economic growth.

Capitalism is grounded in an ideology that pursues unlimited expansion and growth. As such, in a capitalist society, it is natural to prioritize the collection of quantitative data on items that the system desires to increase, such as population, employment, production, sales volume, and wages. Then, the data are inevitably processed, analyzed, and interpreted on the premise that expansion and growth are desirable. Even in the current regional policies, such as Japan's Comprehensive Strategies and the Municipal Version of the Comprehensive Strategies under Overcoming Population Decline and Vitalizing Local Economy, numerical targets have been set primarily by choosing these kinds of data as key performance indicators. On the other hand, the difficult-to-quantify dimensions like quality of life and subjective well-being, as well as the environmental burden, the widening disparities, and the increasing fluidity of employment, which are surely negative side effects of expansion and growth, are not given sufficient attention.

A sense of crisis that capitalism, which pursues relentless expansion and growth, is damaging the natural environment and humanity has led to heightened discussions on the concept of post-capitalism (Gibson-Graham 1996, 2006; Karatani 2003; Harvey 2014; Chatterton and Pusey 2020). Such discussions tend to precede ideology and hence to be utopian, but they are useful for relativizing the notion that there is no hope apart from expansion or growth. Therefore, it is inevitable for data to take on a political nature. However, if we could relativize the expansion-and-growth ideology, we could collect and view data differently than before, enabling us to use data to envision alternative futures that are not entirely pessimistic.

5.4 Population, Immigration, and the Japanese People

Let us suppose that we must stop the decline in population at all costs for the survival of our nation and society. Insofar as the expected results of the existing measures against the declining birthrate have not been achieved, the next procedure to be considered should logically be whether it will be possible to implement full-scale adoption of immigration. The authors of *The Shock of a Shrinking Japan*, however, without mentioning immigration at all, conclude by saying, "Although it may sound irresponsible, we have not found any measure that we can recommend as 'the prescription' and have, therefore, no choice but to give up on presenting one" (p. 195). Whether consciously or unconsciously, the authors seem to have determined that there is no need to mention immigration as a prescription for the "shrinking Japan."

The starting point for *The Shock of a Shrinking Japan* as a project is traced to the "Extinction of Municipalities" (Masuda 2014) and to what may be considered its digest edition, NHK Today's Close-Up "Unipolar Society: The New Population Decline Crisis" (broadcast on May 1, 2014). Masuda (2014) dismissed the possibility of maintaining the population by accepting immigrants, saying, "Large-scale immigration from overseas cannot become a realistic policy as a measure against population decline (p. 92)." The reason he gives for this is that "covering the shortage in the fertility rate would require immigration on a scale that would transform Japan into a multi-ethnic nation, but it is unlikely that a national consensus on this will be reached (p. 92)." Immediately after that, however, he argued that in order to globalize Japan and improve the national productivity amid the declining working-age population, it is necessary to proactively accept "highly skilled human resources" from overseas. He asserted that the technical intern training program should be expanded, particularly in the areas of nursing care and construction, where a serious shortage of human resources is expected in the future (p. 92–93).

These claims by Masuda (2014) are in line with the attitudes and feelings toward foreigners held by the government as well as many Japanese people. In February 2014, the Council on Economic and Fiscal Policy's "Choosing the Future" Committee released a population projection based on the assumption that 200,000 immigrants would be accepted annually (Cabinet Office 2014). This elicited opinions from both inside and outside the committee, which discussed taking advantage of foreigners as a labor force but did not mention immigration. In March 2016, the "Special Committee on Labor Force Assurance" was established within the Liberal Democratic Party. However, the report it submitted (Liberal Democratic Party 2017) concluded that "on the premise of a new framework for their proper management as employed workers, the status of residence for work purposes should be granted after scrutiny of individuals in areas where it is necessary, while giving due consideration to not making mistakes as an immigration policy (including consideration for foreign study and acquisition of qualifications, etc.)." It also added that the period of stay could be extended, but since a prolonged stay would "lead to problems related to invitation of their family members and their resettlement in Japan, further discussions will be needed."

There is a "moral problem" in this attitude: Foreigners are considered necessary as a labor force, but invitation of their family members and their resettlement are considered "problems." This is because foreigners are treated simply as a labor force, i.e., a resource for economic growth, and their rights to form families and to pursue long-term self-fulfillment in Japan are not widely recognized. Many objections to immigration cite the social anxiety and increased costs it would bring about: If this was true, these problems would increase as the foreign population increased, regardless of whether they settle here in Japan. Therefore, this reluctance to accept immigration premised on long-term and permanent residency appears to stem from a different sentiment than feelings about a simple increase in the foreign population.

Perhaps it is rooted in how the line is drawn between foreigners and Japanese people. Since the term "foreigners" refers to people other than Japanese, the criteria for drawing the line are determined by who the Japanese are. There are two definitions of being Japanese: Japanese by nationality and Japanese by ethnicity. If the number of immigrants who live here as long-term or permanent residents increases, people with ethnicity different from the "authentic" Japanese will eventually acquire nationality as Japanese and gain a stronger voice in society, including the right of suffrage—now seen as a threat to the ethnic identity of "authentic" Japanese. Taken a step further, this crisis mentality could dangerously lead to ethnocentrism or racism.

This is not, however, a problem only for Japan. Coleman (2006) described the social change resulting from having the previous majority population born in a low-fertility country being replaced by a high-fertility immigrant population of a different race or ethnicity as the "third demographic transition." In the USA, the population projection that non-Hispanic whites will become a minority by 2043 has sparked debate, resulting in the emergence of the deep-rooted racial consciousness of American society (Lichter 2013). Bialasiewicz (2006) highlighted how two bestsellers published in the USA and Italy both warn that "the West" is under threat as immigrants maintain high fertility rates amid the sluggish birthrate of white-skinned people of Western descent. The "death of the West" described in these books is not merely an allegory of political or geopolitical decline but involves the life and death of real people. Hispanic and Muslim women are believed to participate through childbirth in the recovery of Mexico's lost land from the USA and in the Islamic "Reverse Crusade" in Europe, respectively. Thus, the female body has become "the new battleground for the preservation of the identity of the West" (p. 702).

For many nations, the geopolitics of the population does not only focus on population size and geographical distribution but also on maintaining the implicitly assumed desirable racial and ethnic composition. Since the ideology of encouraging child-bearing is called "natalism," we could use the term "ethno-natalism" to refer to the attitude that prevents people who are different from the host group, in terms of phys-ical features and ethnicity, from settling and reproducing generations in the host country and that aims to maintain and increase the population by recovering the fertility rate of the host group.

5.5 Ordering of the Population

In *The Shock of a Shrinking Japan*, the attitude of treating certain groups as "others" can be seen as directed not only toward foreigners but also toward Japanese people. This is evident in Chap. 1.

While Toshima Ward was recognized by the Japan Policy Council as a "city at risk of extinction," its social dynamics point to a continued excess net migration for many years. However, the majority of the single people who have moved to the ward are low-income earners, such that ward officials described them saying, "Although this is only a generalization, it will be difficult for them to get married and raise children with their annual income level" (p. 25). In addition to their low capacity to pay taxes, the authors said that without getting married to raise future taxpayers, they could lead to an aging population and an increased social security burden. To understand "the work and lifestyle of the single Tokyoites coming from the other regions, who could become a burden to Tokyo in the future" (p. 27), the authors focused on the security sector, which is facing a labor shortage due to the construction rush.

Many of the people who sought employment at security companies were young people from rural areas who were struggling to make ends meet and had come to seek ways to support themselves. After interviewing them in their dormitory room, the authors wrote that "the room was filled with a peculiar mix of the odor of sweat like in a high school athletic club room and the smell of cigarettes" (p. 31). When the authors saw a young man who had moved into the dormitory that day, distributing the local delicacies he had brought to his roommates, they wrote, "It was a strange sight to see that although the room was the size of only six tatami mats, they were greeting each other over their beds, saying, 'I'm ____, from Shizuoka, nice to meet you,' as if they were greeting their neighbors" (p. 32–33).

Upon hearing the stories of dozens of security guards, the book's authors found that the main reason for their moving to Tokyo was that they could not find a suitable job in their hometowns. Elaborating on this, they said that the over-concentration of the population in the Tokyo metropolitan area since 2000 was a "negative concentration" caused by "escaping from the countryside" (p. 39). When the current population of migrants grows older, and the number of young migrants from rural areas decreases, tax revenues will decrease as social security costs increase, making local government finances unsustainable. In addition, if the number of single elderly people increases, the nursing care and medical care systems premised on the support of family members will become unviable. Single elderly persons who have no place to go are already dying alone in hospitals.

Finally, they mention the case of a 67-year-old guard who had been through difficult life experiences and told the authors, "this seems to be the future that awaits these young people" (p. 52). After graduating from university, he moved to Tokyo and worked as a guitarist. In his 50 s, he suffered a stroke and, unable to play anymore, became a security guard. He suffered a stroke again last year and now lives in the dormitory while continuing his rehabilitation. He is living on welfare but aims to return to work. It is not uncommon for security firms to have to support elderly security guards who have no place to go, who have lived alone in dormitories and worked to the limit of their physical strength. Chap. 1 concludes by saying, "These young people can be considered as reserve forces for the 'single elderly persons' who are unable to marry or have a family, and who are only getting older. They may weigh heavily on us in the future as a negative legacy of Tokyo, which had been continuously gathering people all these years" (p. 55).

Here, we see that the working-age population is valued by their ability to pay taxes, while the elderly and the socially vulnerable are regarded as liabilities that only increase social security costs. Low-income single migrants are considered a negative factor that will increase the burden on the Tokyo area in the future by aging without raising the next generation of taxpayers. Thus, the population is being ordered one dimensionally in terms of benefits and costs for the economy and public finances.

Young people from low-income rural areas are being alienated in a truer sense of the word. The authors feel that the young security guards live in a "strange" world that is different from their own. They are, for the authors, "reserve forces for the 'single elderly persons' who are unable to marry or have a family, and who are only getting older" and are, therefore, deemed unproductive members of society. Not only that, but "they may weigh heavily on us in the future as a negative legacy." For the authors, the security guards from the regions are "others" who are different from "us," and the concentration of low-income earners in the Tokyo metropolitan area is nothing more than a "negative concentration."

5.6   Conclusion

Demographic problems facing Japanese society, such as low fertility and aging, are inseparable from geographic problems such as over-concentration in Tokyo and marginal settlements. In other words, the most important policy issues in modern Japan lie in areas where population and geography are closely linked. Population geography has been tasked with analyzing the current state of the population in geography. Population geographers, however, have consciously distanced themselves from debates on ideology and politics, such as what the ideal landscape of population should be and what policy measures should be taken to achieve that ideal. This chapter explored the possibilities of establishing a politico-economic population geography by using *The Shock of a Shrinking Japan* as a guide for discussion.

The author believes that the fundamental question to ask involves the concept of population. The English word "population" and the German word "bevölkerung," established in the era of mercantilism, meant the exercise of authority to "increase the number of people" or to "settle people" (Minami 1957). When a population is defined, analyzed, and discussed by reducing the experiences and facts of each person's life into aggregate quantities, the original meaning of the word surfaces, even today. This is evident in the policy recommendations of the Japan Policy Council and in the logic of the "regional revitalization" promoted by the government (Nakazawa 2016).

Most dictionaries define population as the total number of people living in a country or region. In other words, population is a geographical quantity that has no inherent qualitative differences. However, each individual inevitably senses qualitative differences in population, and this sensed difference unknowingly comes into play along with concomitant value systems in the discourse of population. In light of the concerns about Japan's declining population, the "population" in mind in discussions on policies that should be taken to maintain and increase the population is not the same as the total number of people living in Japan.

Population geography is expanding its horizons from the analysis of populations as homogeneous geographic quantities to the analysis of the geographical diversity of people's lives, the movement of people as lived experiences, and the politics surrounding these factors. Although the importance of quantitative demographic analysis has not completely vanished, it is essential to clearly recognize and deepen the discussion on the fact that the concept of population itself, which defines population geography as an important subdiscipline of geography, is a multidimensional one. Such an understanding will make the ongoing radical transformation of population geography fruitful and meaningful.

References[1]

Bailey AJ (2005) Making population geography. Oxford University Press, New York

Bialasiewicz L (2006) 'The death of the West': Samuel Huntington, Oriana Fallaci and a new 'moral' geopolitics of birth and bodies. Geopolitics 11: 701–724. https://doi.org/10.1080/146 50040600890859

Black J (1998) Maps and politics. University of Chicago Press, Chicago

Cabinet Office (2014) Keizai zaisei shimon kaigi, dai 7 kai kishakaiken no yoshi (Abstract of 7th press interview, Council on Economic and Fiscal Policy). http://www5.cao.go.jp/keizai-shimon/kaigi/special/future/0513/interview.html (J)

Chatterton P, Pusey A (2020) Beyond capitalist enclosure, commodification and alienation: Postcapitalist praxis as commons, social production and useful doing. Prog Hum Geogr 44: 27–48. https://doi.org/10.1177/0309132518821173

Coleman D (2006) Immigration and ethnic change in low-fertility countries: A third demographic transition. Population and Development Review 32: 401–446. https://doi.org/10.1111/j.1728-4457.2006.00131.x

Datta A (2015) New urban utopias of postcolonial India: 'Entrepreneurial urbanization' in Dholera Smart City, Gujarat. Dialogues Hum Geogr 5: 3–22. https://doi.org/10.1177/2043820614565748

---

[1] (J): Written in Japanese.

Gibson-Graham JK (1996) End of capitalism (as we knew it): A feminist critique of political economy. Blackwell, London

Gibson-Graham JK (2006) A postcapitalist politics. Blackwell, London

Giteleman L (ed) (2013) 'Raw data' is an oxymoron. MIT press, Cambridge

Harvey D (2014) Seventeen contradictions and the end of capitalism. Profile Books, London

Headquarters for Overcoming Population Decline and Vitalizing Local Economy in Japan (2015) Machi, hito, shigoto sosei sogo senryaku (Comprehensive strategy for overcoming population decline and vitalizing local economy). https://www.chisou.go.jp/sousei/info/pdf/h27-12-24-sir you2.pdf. Accessed 1 Oct 2022 (J)

Karatani K (2003) Transcritique: On Kant and Marx. MIT press, Cambridge

Kitchen R (2013) Big data and human geography: Opportunity, challenges and risks. Dialogues Hum Geogr 3: 262–267. https://doi.org/10.1177/2043820613513388

Kitchen R (2014) The real-time city? big data and smart urbanism. GeoJournal 79: 1–14. https://doi.org/10.1007/s10708-013-9516-8

Kwan MP (2002) Feminism visualization: Re-envisioning GIS as a method in feminist geographic research. Ann Assoc Am Geogr 92: 645–661. https://doi.org/10.1111/1467-8306.00309

Liberal Democratic Party (2017) Kyosei no jidai ni muketa gaikokujin roudousha ukeire no kihonteki kangaekata (Basic guideline of introducing foreign workers in the age of multiculturalism) https://www.jimin.jp/news/policy/132325.html

Lichter DT (2013) Integration or fragmentation? Racial diversity and the American future. Demography, 50: 350–391. https://doi.org/10.1007/s13524-013-0197-1

Luque-Ayala A, Marvin G (2015) Developing a critical understanding of smart urbanism? Urban Stud 52: 2105–2116. https://doi.org/10.1177/0042098015577319

Masuda H (ed) (2014) Chiho shometsu: Tokyo ikkyoku shuchu ga maneku jinko kyugen (Disappearance of peripheral regions: Rapid population decline driven by the predominance of Tokyo). Chuokoron-Shinsha, Tokyo (J)

McFarlane C, Söderström O (2017) On alternative smart cities: From a technology-intensive to a knowledge-intensive smart urbanism. City 21: 312–328. https://doi.org/10.1080/13604813.2017.1327166

Minami R (1957) Jinko no gainen (The concept of population). In: Shimonaka Y (ed) Jinko daijiten (Population encyclopedia). Heibonsha, Tokyo, p 3–7 (J)

NHK Special crew (2017) Shukusho nippon no shogeki (The shock of a shrinking Japan). Kodansha, Tokyo (J)

Nakazawa T (2016) Chihososei no mokutekiron (Teleology of 'regional revitalization'). Keizai Chirigaku Nenpo (Annals of the Japan Association of Economic Geographers) 62: 285–305. https://doi.org/10.20592/jaeg.62.4_285 (J)

Peck J (2002) Political economies of scale: Fast policy, interscalar relations, and neoliberal workfare. Econ Geogr 78: 331–360. https://doi.org/10.1111/j.1944-8287.2002.tb00190.x

Population Decline Examination Subcommittee, Japan Policy Council (2014) Seicho wo tsuzukeru nijuisseiki no tame ni: Sutoppu shoshika, chiho genki senryaku) (For sustaining growth in 21st century: Strategies for overcoming low fertility and vitalizing local economy). http://www.pol icycouncil.jp/pdf/prop03/prop03.pdf. Accessed 1 Oct 2022 (J)

Robbins P, Smith SH (2017) Baby bust: Towards political demography. Prog Hum Geogr 41: 199–219. https://doi.org/10.1177/0309132516633321

Seager J (2003) The atlas of women: An economic, social and political survey. Women's Press, Toronto

Sheppard ES (1993) Automated geography: What kind of geography for what kind of society? Prof Geogr 45: 457–460. https://doi.org/10.1111/j.0033-0124.1993.00457.x

Sieber R (2006) Public participation geographic information systems: A literature review and framework. Ann Assoc Am Geogr 96: 491–507.https://doi.org/10.1111/j.1467-8306.2006.007 02.x

Taylor PJ (1990) GKS. Polit Geogr Q 9: 211–212.https://doi.org/10.1016/0260-9827(90)90023-4

Taylor PJ, Overton M (1991) Further thought on geography and GIS. Environ Plan A 23: 1087–1090.https://doi.org/10.1068/a231087